EMBROIDERY · MADE EASY

BEAUTIFUL
Birds

BETH HOYES

Brimming with creative inspiration, how-to projects, and useful information to enrich your everyday life, quarto.com is a favorite destination for those pursuing their interests and passions.

© 2022 Quarto Publishing Group USA Inc.
Artwork and text © 2022 Beth Hoyes

First published in 2022 by Walter Foster Publishing, an imprint of The Quarto Group.
100 Cummings Center, Suite 265D, Beverly, MA 01915, USA.
T (978) 282-9590 **F** (978) 283-2742 **www.quarto.com** • **www.walterfoster.com**

Walter Foster Publishing titles are also available at discount for retail, wholesale, promotional, and bulk purchase. For details, contact the Special Sales Manager by email at specialsales@quarto.com or by mail at The Quarto Group, Attn: Special Sales Manager, 100 Cummings Center, Suite 265D, Beverly, MA 01915, USA.

ISBN: 978-0-7603-7536-5

Digital edition published in 2022
eISBN: 978-0-7603-7537-2

Freelance project editor: Stephanie Carbajal

Printed in China
10 9 8 7 6 5 4 3 2 1

Table of Contents

Introduction 5

Tools & Materials 6

Stitching Techniques 8

Finishing Touches 17

Step-by-Step Projects

    Calliope Hummingbird 18

    Cardinal 26

    Keen-Billed Toucan 34

    Wood Duck 42

    American Flamingo 50

    Kingfisher 58

    Atlantic Puffin 66

    Swallow 74

    Barn Owl 82

    Hoopoe 90

    Raven... 98

    Eastern Bluebird 106

Pattern Templates 114

About the Author 128

INTRODUCTION

Welcome to Beautiful Birds!

I'm so excited to have you along on this journey while we stitch birds from across the world. This book is all about making the detailed and expressive process of thread painting more accessible, while simultaneously celebrating so many lovely bird species. Birds and feathers are one of my favorite things to stitch, and I'm sure you'll enjoy these patterns too. There's a mix of portraits, close-ups, and birds in flight—so there's a lot of different learning to do as you meet these birds through stitching.

Birds are full of different textures, character, and detail, so I've done my best to capture them all in their true dapperness. If you're new to stitching, this book walks you through the basics and guides you with each pattern step by step, including in-depth color guides and photos of different stages. If you're a stitching pro, enjoy some new patterns and add your own flare wherever you fancy. This book is all about taking it slow and working through each pattern a few stitches at a time, making it as easy to follow as possible and giving you the tools to start feeling more confident with creating your own intricate nature embroideries.

Let's get started!

Tools & Materials

For each pattern, you'll need an 8" x 8" inch (or larger) square of white, cream, or light-colored fabric. I often find muslin and linen to be the best types of fabric to work with. You'll need sharp scissors to cut your fabric and thread, embroidery thread (refer to the color key in each pattern for color palettes; color reference numbers are based on DMC® embroidery threads), an embroidery needle, an embroidery hoop, and a well-lit, comfy spot. These patterns are designed for 6-inch embroidery hoops, but you can use larger hoops if you wish. When sourcing thread, look for six-strand embroidery thread, ideally made by DMC. For embroidery needles, I recommend size 1, but smaller or larger needles will work, if preferred.

For these patterns you'll need to use thin thread, which can be easily made by dividing the six-strand thread into three parts of two strands each. To do this, take a length of embroidery thread (usually about an arm's length) and rub one end between your index finger and thumb until it separates into six strands, and then pull two strands away from the rest to stitch with. Repeat this until you have three sets of two strands. The threads should separate easily, but if not, try again with a shorter length of thread.

Transferring Patterns

To transfer a pattern, choose the bird you want to stitch from the pattern section at the back of this book, and cut that pattern out. Lay the fabric over the pattern and check its positioning to make sure the bird will be in the right place within the hoop. For these patterns, it often works best to place the bird in the center of the hoop. Attach the fabric to your pattern either with masking tape or a couple of pins at the top and bottom. With one hand, hold the fabric and pattern still and firm against a bright, sunny window or light box; with the other, use a sharp, soft (B–3B) pencil, thin pen, or washable fabric pen to trace the outline of the pattern. Aim to get the main details sketched onto your fabric, like the outlines and main lines of the template; you can add the smaller details later if it's easier. When you've finished tracing the pattern onto the fabric, take your fabric off the paper pattern template. If your pattern is faint or not clearly visible, place the fabric on a flat surface and go over the pattern with a pencil or pen to make the lines more visible and/or sharper. Once you're happy with your pattern template, load your fabric into the hoop (to learn how, see the next page). You can reuse your paper pattern template as many times as you like.

Setting Up Your Fabric & Hoop

When you've transferred your pattern onto the fabric, separate your hoop into two parts that look like rings, one without the screw and one with the screw. Place the ring without the screw on a flat surface, and then place your fabric on top of it. At this point (once you have transferred your pattern), you can roughly place the fabric in a way that captures the pattern where you want it in the circular frame of the hoop. Place the second ring of the hoop with the screw on top of the fabric and other ring, so that the fabric fits snugly between the hoops.

You want the fabric to be taut in the hoops and stretched flat. If there are wrinkles in the fabric, you can loosen the screw slightly and pull at the skirt of fabric outside of the hoop near the wrinkle until it is flat. You want at least 1 inch of fabric-skirt around the outside of the hoop. If you like, you can trim the fabric around the hoop into a circular shape before or after loading it in the hoop, so it's easier to stitch.

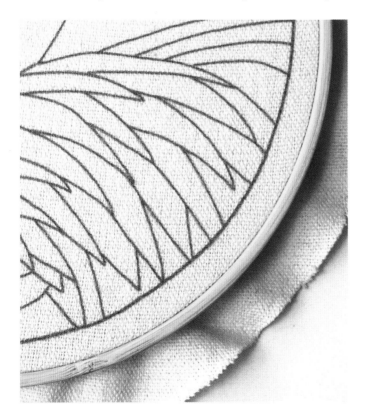

At this point, you may want to adjust the fabric placement by tightening and/or loosening the screw until you have a good, tight fit and placement. For most of the patterns in this book I recommend placing the bird in the center of your hoop. For the close-up patterns, like the flamingo, align the circular edge of the pattern with the edge of the hoop, as shown here.

Stitching Techniques

Blending Colors

Where there are several colors in an area of the pattern, you can blend stitches to create color gradients. When blending colors in smaller parts of a pattern where there are two colors, you can create a stripe-like effect by alternating the colors of your stitches with rows of teeny stitches, like in the hummingbird pattern on page 116. These images show the stages of using the pattern color guide to see where the combined colors are before stitching these stripe-like parts toward the hummingbird's shoulder.

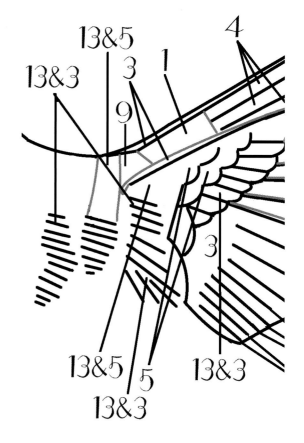

When blending colors, it's easiest to start with one color, leaving gaps between your stitches for the second color. Leave bigger gaps where there are three colors in an area of the pattern, and alternate between adding different stitch colors across the area you're stitching until you are happy with the balance. When you want one color to be more dominant, add more stitches of that color. If you want to blend even further, you can make the thread thinner by using the same separating process as with two-strand thread, pulling one strand away from the rest, instead of two. Then, with the single-strand thread, add more stitches of the color in the area where you want more of it, in the same way you might add dabs of different colors when painting. Thread painting works in layers, so the great news is, you can keep adding layers of color until your heart is content. Shown is another example of blending with a larger area in the wood duck pattern.

STRAIGHT STITCH

Straight stitches are the simplest form of stitches; you start by threading your needle and thread up through your fabric where you want the stitch to start, returning back down through the fabric where you want the stitch to end. Once you have finished your first stitch, you can then repeat this process where you want the next stitch to be, leaving gaps between stitches or not, depending on the effect you want to create. Straight stitches can be stitched in any direction on your embroidery. This stitch is great for creating fur or feathers and can be used in a spontaneous way to cover an area or in an organized way to stitch small patterns.

RUNNING STITCH

Running stitch is a super simple stitch. You simply make a row of short stitches, creating a dashed line across your embroidery fabric. This stitch is also used for gathering the back of your fabric around your hoop when you've finished stitching (see "Finishing Touches" on page 17).

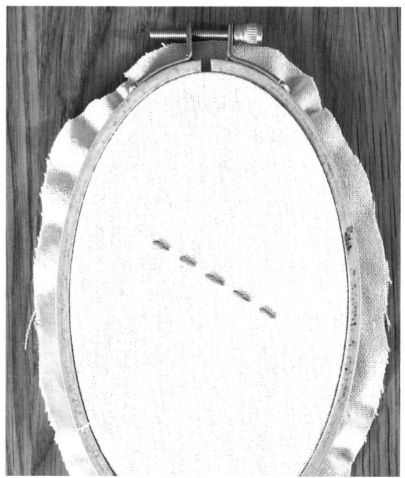

FRENCH KNOTS

Some of the patterns in this book require French knots. French knots can be used to create dotlike patterns with knotted stitches. Start by threading your needle and thread up through the fabric where you want the French knot to be with one hand. Then, with your other hand, hold the thread taut above your fabric. With your sewing hand, loop the thread around your needle once before returning your needle and thread back at the point where you came up through your fabric. For true French knots, you wind the thread around your needle twice, but I often find once is easiest at first and works for these patterns. Let the thread run slowly through your finger and thumb as you pull the needle through the fabric with your sewing hand, allowing a knot to form on the fabric surface.

SATIN STITCH

Satin stitches can be used when you want a smooth area of one color. To satin stitch, start with a straight stitch and then with the following stitch, come up through the fabric right next to the beginning of your first stitch, repeating this process until you've covered the section you're working on. You are aiming to create a row of stitches that are nice and snug, with no gaps.

Staged & Scattered Stitching

Many of these patterns ask for staged or scattered stitches. Staged stitching refers to a series of short rows of stitches on top of each other. I refer to this way of stitching as "staged stitching" because you work in stages, adding rows as you go and creating a sense of the short feathers often found on birds' chests, necks, and bellies. As you stitch your way down the pattern in short rows, follow the stitching flow directions for the pattern you're working on. These photos show this process, as well as how you can sketch lines onto your fabric to guide you as you go. Staged stitching is used for the back of the toucan's head, neck, and chest.

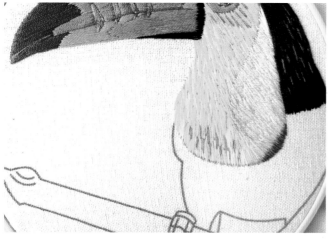 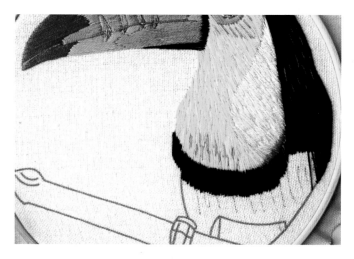

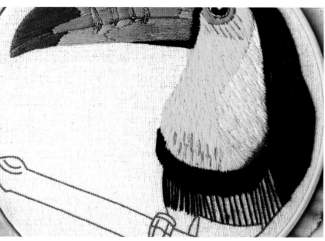 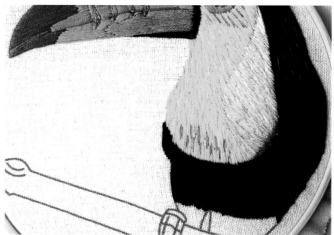

Scattered stitching involves lots of small stitches stitched randomly in relation to each other, rather than close together or in rows. Use the stitching flow directions to guide you with rough directions to aim for as you scatter your stitches. Imagine you are throwing seeds out to birds, with the seeds falling in a scattered way. These stitches work well when blending colors as well, creating a more natural gradient of color in feather-like patterns rather than rows.

Scattered stitching is used to blend cream and white stitches on the puffin's belly.

Tips for Stitching Birds

All birds have similar features, and practicing these features for one bird can help with stitching others. Let's take a look at some of these similar features.

EYE-RING

Many birds have rings of color or thin markings of defining color around their eyes, known as the eye-ring. The color or colors of these vary, but you can use similar series of stitches for each bird. Use the photos and color guides in each pattern to guide you with these small details. For more defined, circular eye-rings, like in the raven pattern shown here, start by following the outer circle of the eye with a series of very short stitches. When you have stitched the first series of stitches, add a series of short stitches that overlap the first set, with each starting and ending in the middle of your first stitches. Overlapping stitches in this way creates a smoother line for the circle.

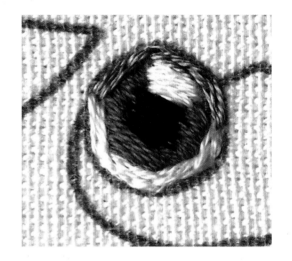

FEATHERS

Feathers are one of the coolest things about birds, but they can seem tricky to stitch at first, with all the texture and different directions. In this book you'll find stitching flow directions diagrams for each pattern; use these to guide you with which angle to stitch as you stitch the different areas of the bird's feathers. For the shorter, soft feathers on birds' necks, chests, and bellies, create series of short rows of stitches or short scattered stitches. For larger feathers, imagine how a real feather looks, with a central bone-like spine, called the "rachis" or "shaft," where super-thin fibers of the feather grow outward

diagonally on either side in branch-like formations. For many of the larger feathers in these bird patterns, you can use straight or diagonal series of satin stitches, referring to the stitching flow diagrams to guide you. The raven pattern is a great example of the central structure in feathers and how to use satin stitch for feathers.

WINGS

Finally, almost all birds all have wings. The texture and color of bird wings vary quite a bit, but the essential structure of the wings is often similar. Birds' wings are often made up of interlocking thinner feathers that can be seen more clearly when birds are in flight. For outstretched wings that are in mid-flight, it can help to create lines across the wings that outline the structure of feathers before filling in the stitches that make up the feather color or texture. The pattern instructions will guide you through these stitches in stages, usually stitching one wing first before the other. You can see this process of stitching the upper-wing structure first in this photo of the Eastern bluebird pattern.

Using the Color Guides & Stitching Flow Diagrams

In this book, you'll find four color guides for each pattern, which correspond to different parts of the bird you'll be stitching, and a color key to guide you. These guides work in a color-by-numbers style—the numbers in the color key correspond with the thread colors you'll be using and show you where to stitch which colors on your embroidery. For each pattern, use the color guides as you stitch. The black lines on the color guides represent the lines of the pattern already marked on your pattern template, and the dark gray lines show you where there is a change in color that isn't marked on the pattern template. When stitching different parts of the pattern, you can sketch these gray lines onto your fabric as you go if it is helpful.

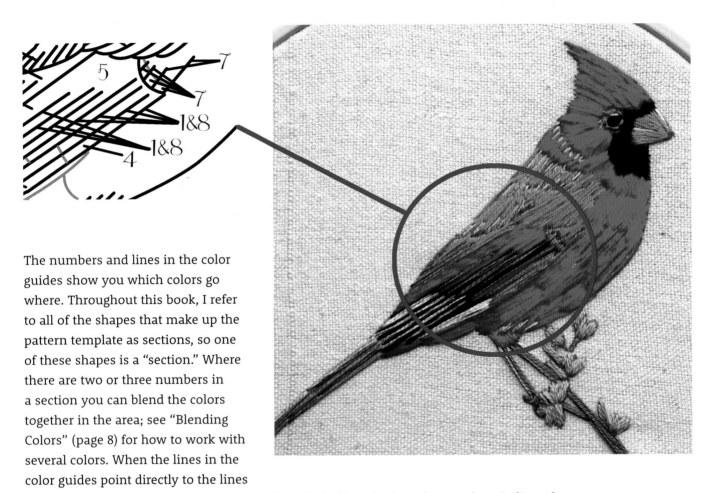

The numbers and lines in the color guides show you which colors go where. Throughout this book, I refer to all of the shapes that make up the pattern template as sections, so one of these shapes is a "section." Where there are two or three numbers in a section you can blend the colors together in the area; see "Blending Colors" (page 8) for how to work with several colors. When the lines in the color guides point directly to the lines in the pattern, as shown above, you should stitch the lines in the color numbers indicated.

For each pattern you'll find a stitching flow diagram, with the flow of stitches marked with blue arrows. As you stitch each part of the pattern, refer back to the diagram and angle your stitches to match the directions of the arrows. This doesn't have to be exact, but the diagrams give you an idea of the stitching angles to aim for; the arrows in the stitching flow diagrams mirror the natural flow of feathers for each bird in real life. For patterns with lots of smaller sections that make up larger sections, there may be different angles to work on close together—just take each part step by step, and you'll be stitching like a pro in no time.

Finishing Touches

When you've finished your embroidery, tidy up the fabric at the back of your hoop. Start by chopping off any loose ends, without cutting your knots. For any loose or baggy stitches, hook the baggy stitch at the back of your embroidery with your needle, before fixing it down with a few stitches on the back of the embroidery. To gather the back of your fabric around your hoop, turn your finished embroidery over. Take a length of thread and make a double knot at one end of the thread and thread your needle through the top of the skirt of fabric around the hoop near the screw, about ½ inch away from the edge of your hoop. Then create a row of short running stitches all the way around the edge of your hoop, staying about ½ inch away from the edge of the hoop. When you reach the knot where you started, pull your thread tight, so that the fabric ruffles up into the middle of the back of your hoop. Then stitch a few small stitches where you finished, near the knot, to hold the thread tight. Note that in the diagram I've used blue thread so it's easy to see; however, generally you want to use a color that matches the fabric as closely as possible so that it doesn't show through.

This will add a neat finish to your hoop and a smooth line if you want to hang your finished piece on the wall. If you want to turn your bird embroidery into a patch or keep it in a square fabric shape, you can unscrew your hoop and remove the embroidery before turning it into whatever you fancy.

Calliope Hummingbird

Hummingbirds are speedy little birds found flying around many mountain and tropical regions. The calliope hummingbird is the smallest bird in the United States. Hummingbirds have many curved feathers, creating little shell-like patterns. As hummingbird feathers are so small and shiny, they often light up in myriad colors, so this is a particularly colorful pattern. It may feel a little fiddly in parts, but it's so worth it to get the full-jeweled beauty of these amazing little creatures!

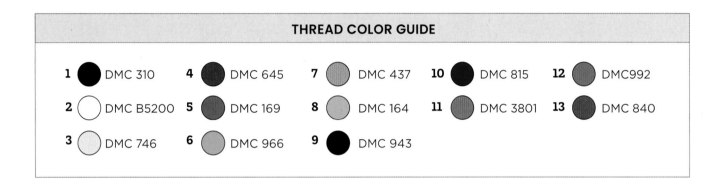

	THREAD COLOR GUIDE								
1	DMC 310	4	DMC 645	7	DMC 437	10	DMC 815	12	DMC992
2	DMC B5200	5	DMC 169	8	DMC 164	11	DMC 3801	13	DMC 840
3	DMC 746	6	DMC 966	9	DMC 943				

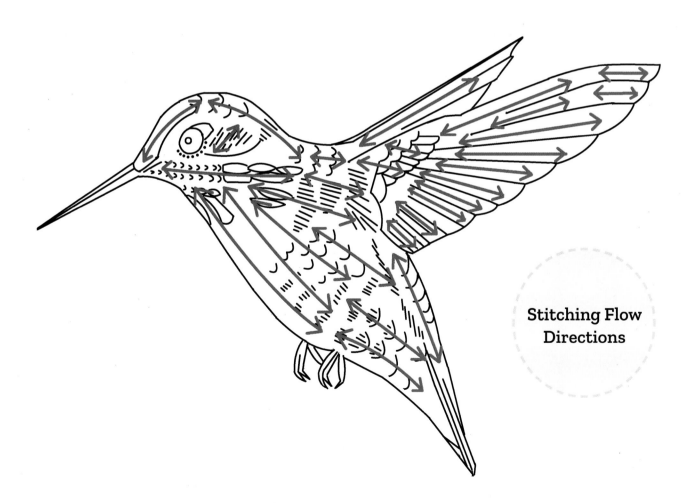

Stitching Flow Directions

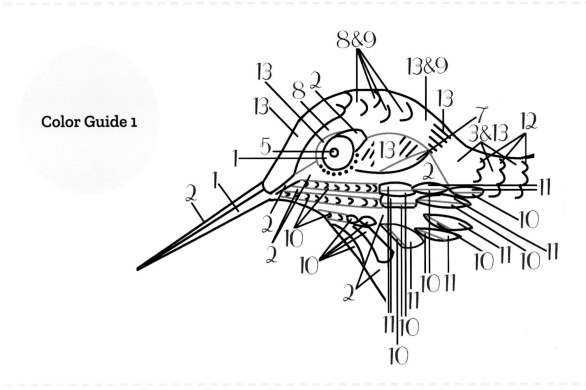

Color Guide 1

TIP

The hummingbird's face may feel like the fiddliest bit on this pattern. Just take it one step at a time, and you'll be done before you know it!

STEP 1

Start by stitching the hummingbird's face, beak, and eyes, using Color Guide 1 for reference. The feathers on the hummingbird's face are very small, so use staged stitching (see page 12) as you work on various parts of the pattern. For the dotted markings under the eyes, you just need a couple of teeny, stacked stitches for each dot in Light Pewter (169). Then, in between these dotted markings, stitch another row of dots in Very Dark Beaver Gray (645). Follow the gray lines in Color Guide 1 as you add the smaller details to the hummingbird's face, such the Medium Beige Brown (840) line above the hummingbird's eye, and where you change colors, such as between the Snow White (B5200) and Light Forest Green (164) sections around the eyes.

STEP 2

Finish this section by stitching the top of the hummingbird's chest, using staged stitching in Snow White (B5200).

STEP 3

Move on to the hummingbird's shoulders and wings. Start with the shoulders and the inner areas of the wings before working out to the upper wings, following Color Guide 2 (see page 22). The wings are made up of striped markings; it's easiest to begin with the thinner stripes first. Start with the Black (310) lines toward the bottom of the top wing, following the lines on the template. Add the longer black lines before filling in the sections between these lines.

Color Guide 2

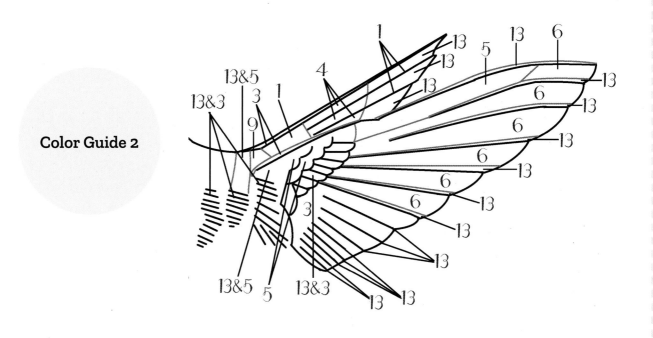

STEP 4
Follow the gray lines in Color Guide 2 that indicate where to change colors as you stitch the wings. For the lower wing, begin with single stitches in Medium Beige Brown (840), following the pattern template.

STEP 5
Then, as you move toward the top of the wing, make these brown lines thicker, as shown in Color Guide 2. Wahoo! You're nearly there on the wings; now you can work on the Medium Baby Green (966) and Off White (746) sections between the lines.

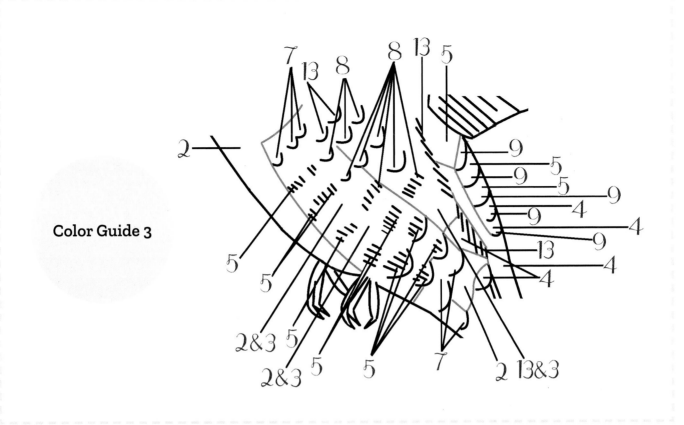

Color Guide 3

STEP 6
The hummingbird's chest and back are made up of many curved and dashed markings. It helps to stitch the smaller, curved shapes first, before filling in the spaces between them.

STEP 7
When you have finished the curved and dashed markings, start working on the larger areas of color between them, referring to the gray lines in Color Guide 3 that indicate where to place the different sections of color.

Color Guide 4

STEP 8

All that's left are the hummingbird's tail and feet. Use Color Guide 4 to help with placement of colors on the tail. As with the wings, start with the thinner lines of Light Tan (437) on the pattern template before adding the Black (310) markings on either side and the tip of the hummingbird's tail. Finally, add the Snow White (B5200) section that makes the base of the hummingbird's tail.

STEP 9

Finish stitching the hummingbird's feet, which are just made up of a series of teeny straight stitches, following the shape on the pattern template in Light Pewter (169), with even teenier Black (310) stitches to finish the claws.

Cardinal

As the state bird of several states, cardinals are well-recognized in the United States. They are often spotted as beautiful flashes of scarlet red in woodland, thickets, suburban gardens, and deserts.

THREAD COLOR GUIDE					
1 DMC 310	4 DMC 351	7 DMC 169	9 DMC 355	11 DMC 840	
2 DMC 746	5 DMC 666	8 DMC 334	10 DMC 164	12 DMC B5200	
3 DMC 350	6 DMC 3743				

Stitching Flow Directions

Color Guide 1

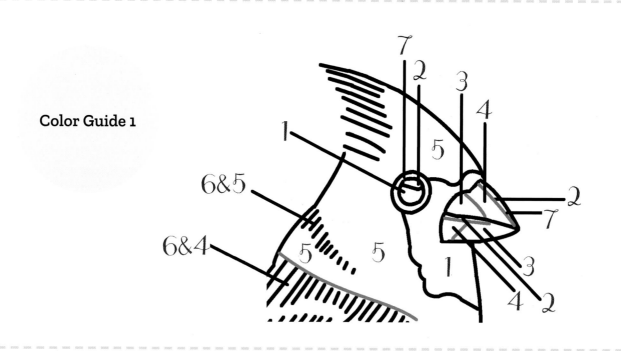

STEP 1

Start by stitching the cardinal's eyes and the Black (310) area around the beak. Refer to page 14 for tips on stitching birds' eyes as you stitch the Light Pewter (169) outline around the cardinal's eyes. Next, work on the cardinal's beak. There are two Off White (746) lines to add, one covering most of the beak's centerline and the other about two-thirds of the way down the top line of the beak. After adding these lines, add a very short stitch in Light Pewter (169) near the upper tip of the beak. Then refer to the gray lines in Color Guide 1 as you stitch the Medium Coral (350) and Coral (351) sections.

STEP 2

Finish the cardinal's face and dapper head tuft with the Bright Red (666) stitches. You can also start stitching the first of the Very Light Antique Violet (3743) dashed markings before filling in the Bright Red (666). You may find it easiest to start by stitching all the Very Light Antique Violet (3743) dashed stitches on the cardinal's neck and wings before filling in the stitches around them.

STEP 3

The wings are made up of lots of striped markings, which means lots of skinny stitches, following the pattern template. Begin with the Very Light Antique Violet (3743) dashed stitches before filling in the Coral (351) stitches between and around these on the upper wings. Next, add the Bright Red (666) markings on the mid-section of the wings before filling in the Coral (351) stitches between these stripes.

STEP 4

Stitch the Light Pewter (169) lines on the mid-back section of the cardinal's wings and the small curved shape near the chest before stitching the Bright Red (666) stitches in between.

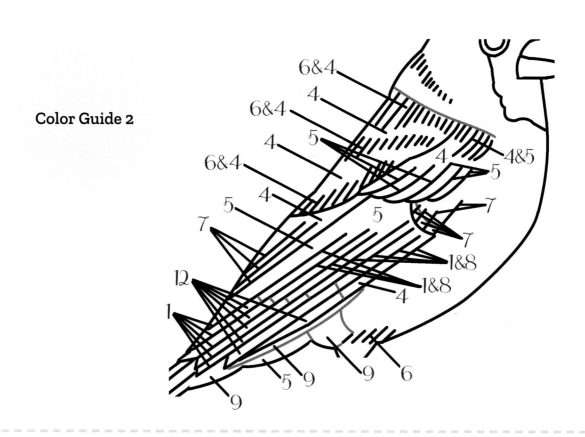

Color Guide 2

6&4
4
6&4
4
5
6&4
4
5
7
12
1
9
5
9
9
6
4
4
1&8
1&8
7
7
7
5
4&5
5
4

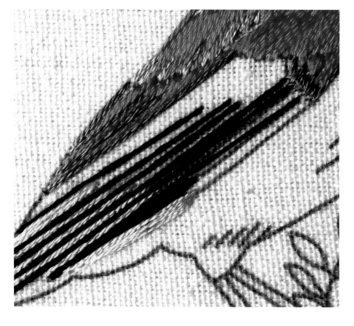

STEP 5

For the next set of wing stripes, start with the Black (310) lines across the bottom of the wings and the shorter Snow White (B5200) stitches in between the black stitches. Under each of your black line stitches, add the Medium Baby Blue (334) stitches, starting at the top of the white stripe stitches and ending at the top of the black stitches.

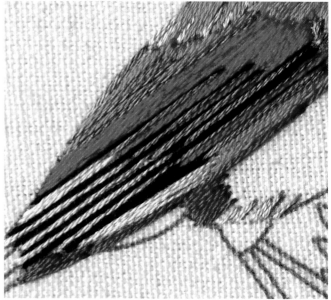

STEP 6

Fill in the Bright Red (666) stitches and add the smaller Dark Terra Cotta (355) markings, shown with the gray lines in Color Guide 2. Finally, stitch the Very Light Antique Violet (3743) dashed markings.

Color Guide 3a

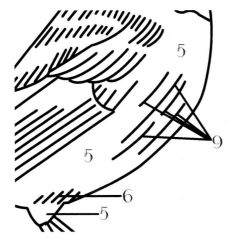

Color Guide 3b

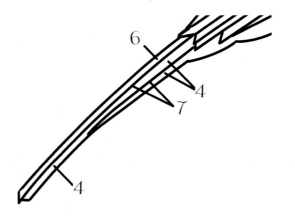

STEP 7

You are ready to start stitching the cardinal's chest. Start with the Dark Terra Cotta (355) lines marked on your template and in Color Guide 3a before filling in the Bright Red (666) stitches around them using staged stitching.

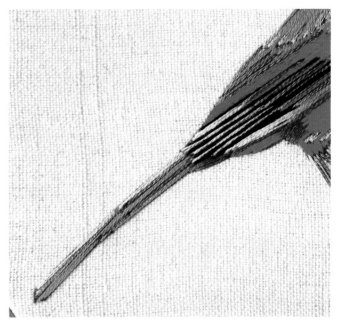

STEP 8

Next, stitch the cardinal's tail, using Color Guide 3b to help place the colors. Start with the Light Pewter (169) stripes and Very Light Antique Violet (3743) stitches before filling in the gaps with the Coral (351).

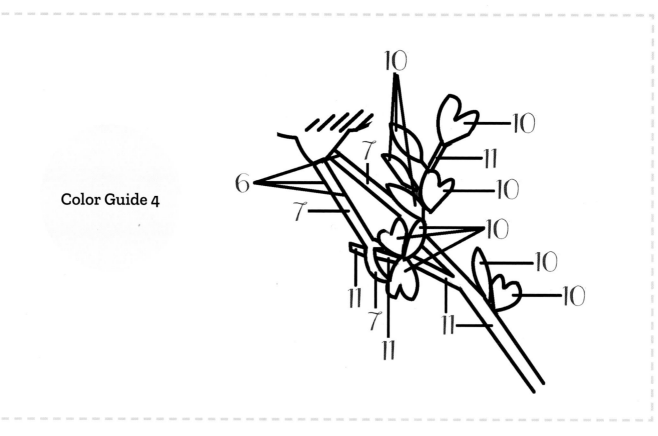

Color Guide 4

STEP 9

Lastly, stitch the cardinal's legs and the branch and leaves. Start with the legs, which are mostly Light Pewter (169) with Very Light Antique Violet (3743) thin lines along the inner and outer edges of the left leg and the outer edge of the right leg. Use Color Guide 4 to help you place these stitches. Use several teeny stitches, following the shape of the cardinal's feet. Then stitch the Medium Beige Brown (840) branches and the Light Forest Green (164) leaves. For the leaves, use a series of satin stitches following the shape of each leaf.

Keen-Billed Toucan

Toucans are known for their beautiful, large, colorful beaks. The keen-billed toucan has a particularly lovely beak of chartreuse green, orange, turquoise, and red. The keen-billed toucan is found in rainforest climates across southern Mexico and South America. This pattern is a closeup of this lovely bird, so the vibrant colors can stand out in your embroidery.

THREAD COLOR GUIDE				
1 DMC 310	4 DMC 703	7 DMC 3853	10 DMC 445	13 DMC 840
2 DMC B5200	5 DMC 704	8 DMC 3846	11 DMC 334	14 DMC 355
3 DMC 746	6 DMC 169	9 DMC 307	12 DMC 869	

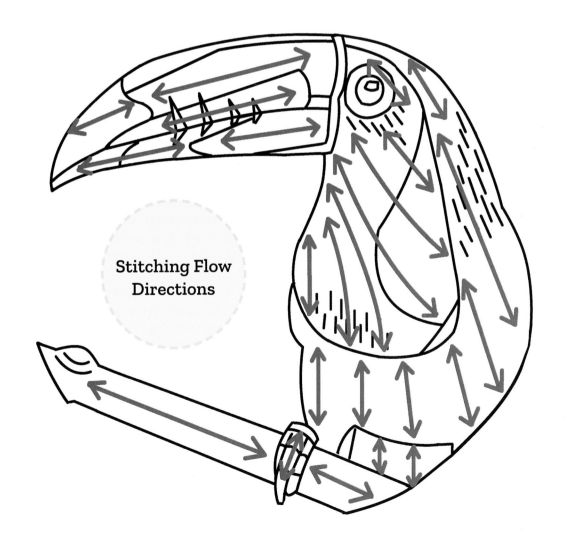

Stitching Flow Directions

Color Guide 1

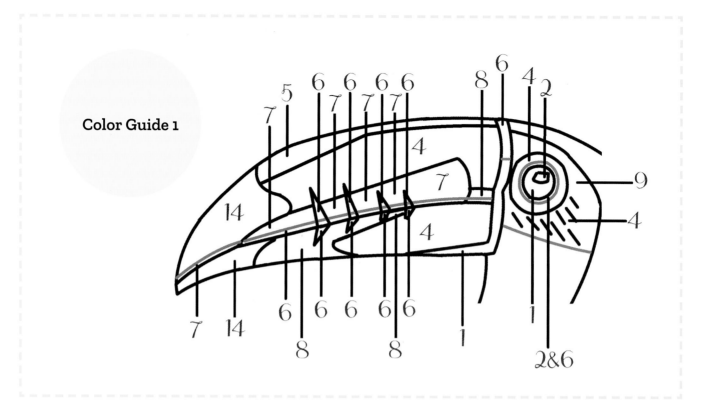

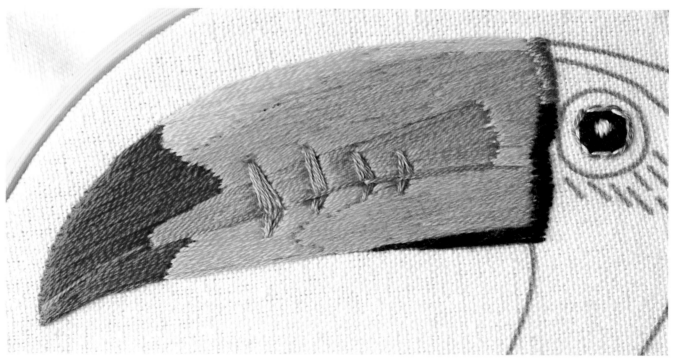

STEP 1

For the toucan's beak, following Color Guide 1, start with the larger color sections before working on the smaller sections. You can stitch the triangular shapes on the beak vertically, rather than following the stitching flow directions. When you've finished the main beak sections, refer to the gray lines in Color Guide 1 as you add the thin lines of Dark Autumn Gold (3853) in the middle tip of the toucan's beak and the Light Pewter (169) line along the middle of the beak. Next, work on the toucan's eye. Start with the central Snow White (B5200) and Black (310) sections in the center of the eye before adding the inner Snow White (B5200) line around your toucan's eye and the outer Light Pewter (169) line stitched closely around the first line (see page 14 for help with stitching the eye-ring).

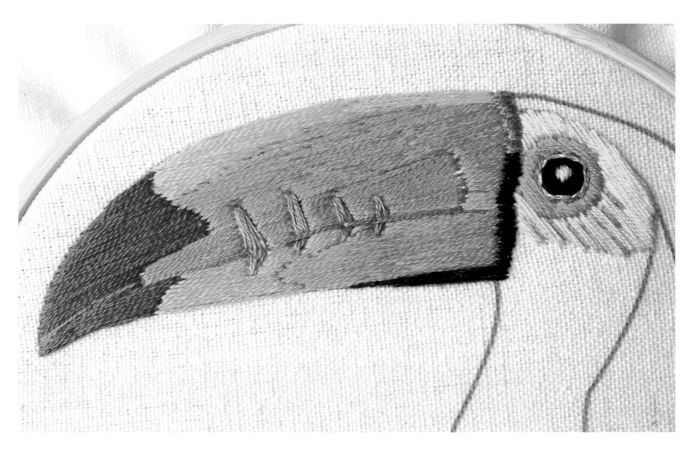

STEP 2

Finish the outer circular section in Chartreuse (703) around the eye. Now work on the Chartreuse (703) dashed markings before filling in the Lemon (307) section around them, referring to the gray line in Color Guide 1 for where this color section ends.

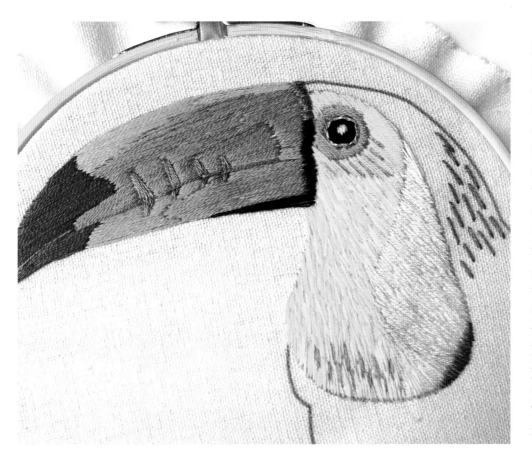

STEP 3

Start on the toucan's chest and back. The chest requires a lot of blending (see page 8); use Color Guide 2 on the following page to help you place the different coloring. It's easiest to start with the Chartreuse (703) dashed markings toward the bottom of the chest before working on the larger section. When you've finished the chest, stitch the Dark Terracotta (355) dashed markings on the back.

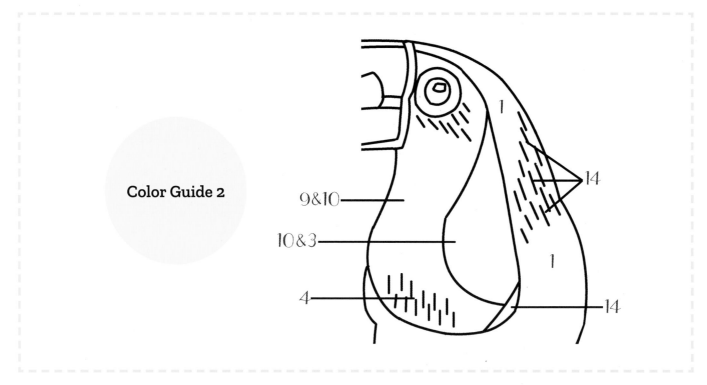

Color Guide 2

9&10

10&3

4

14

1

1

14

STEP 4

Add the Black (310) stitches around the dashed markings. For this part, use staged stitching (see page 12). Now the toucan's colorful beak and chest are stitched, and you've got a good start on the back.

STEP 5

Follow Color Guide 3 on the following page as you continue to stitch the toucan's back and lower belly—you may find it easier to place some spread-out stitches along the contours of the toucan's shape to guide you with the staged stitching.

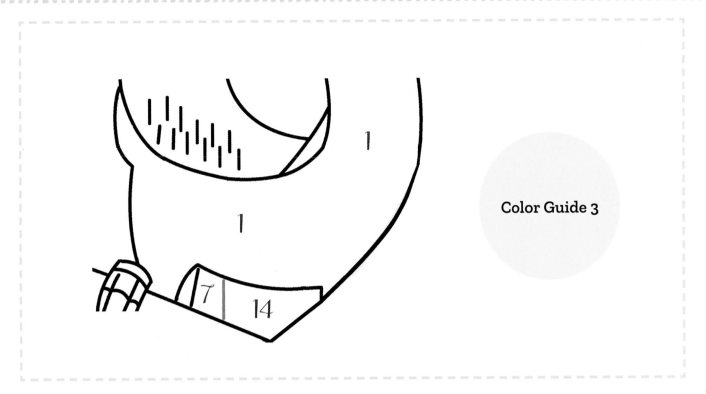

Color Guide 3

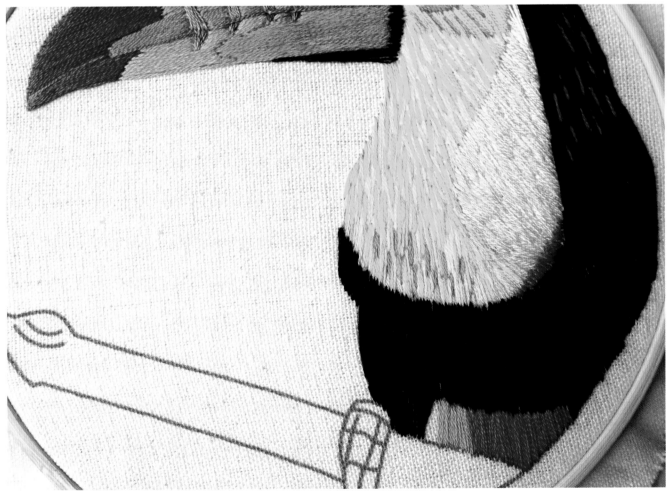

STEP 6

Once you finish the toucan's belly, stitch the leg before adding the Dark Terra Cotta (355) and Dark Autumn Gold (3853) tail sections.

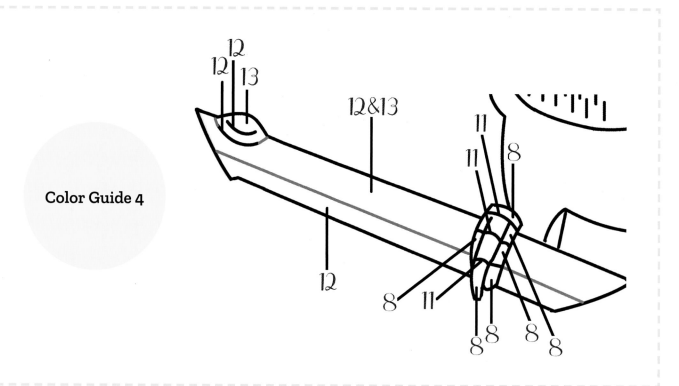

Color Guide 4

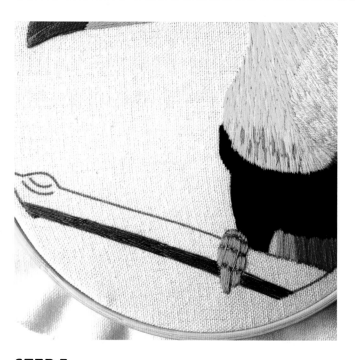

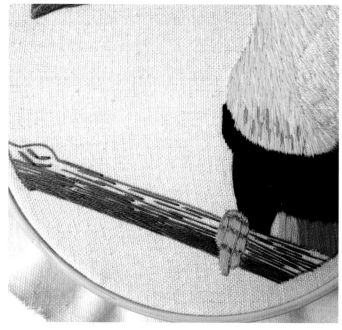

STEP 7

You are almost done—just the foot and branch to go! To stitch the foot, begin with the Medium Baby Blue (334) outlines before stitching the Light Bright Turquoise (3846) sections that make up the rest of the foot. Then it's time to stitch the branch the toucan is sitting on. Start with the Very Dark Hazelnut Brown (869) part of the branch, referring to the gray line in Color Guide 4 to guide you.

STEP 8

For the blended upper section of the branch and the small knob on the branch, start with the Very Dark Hazelnut Brown (869) curved outlines before filling in the rest of the shape.

Wood Duck

The wood duck is one of the most colorful ducks in North America, usually found in watery places like lakes, ponds, and rivers. Due to the colorful markings of this duck, this pattern is super colorful.

THREAD COLOR GUIDE

1 ● DMC 310	5 ● DMC 351	8 ● DMC 158	11 ● DMC 701	14 ● DMC 991
2 ○ DMC B5200	6 ● DMC 743	9 ● DMC 437	12 ● DMC 890	15 ● DMC 992
3 ● DMC 169	7 ● DMC 666	10 ● DMC 3041	13 ● DMC 924	16 ● DMC 400
4 ● DMC 645				

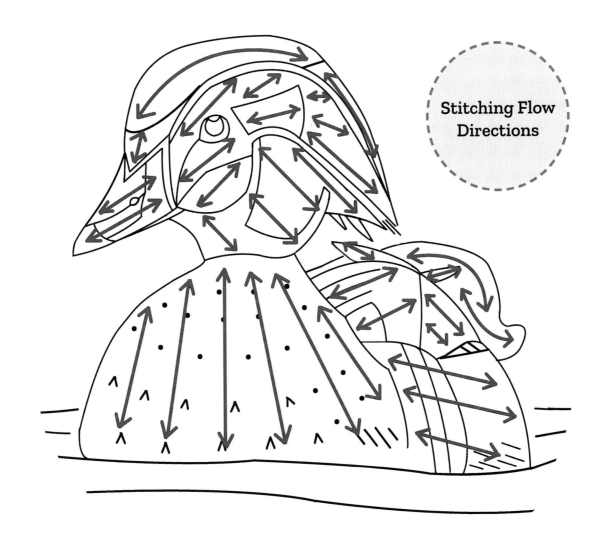

Stitching Flow Directions

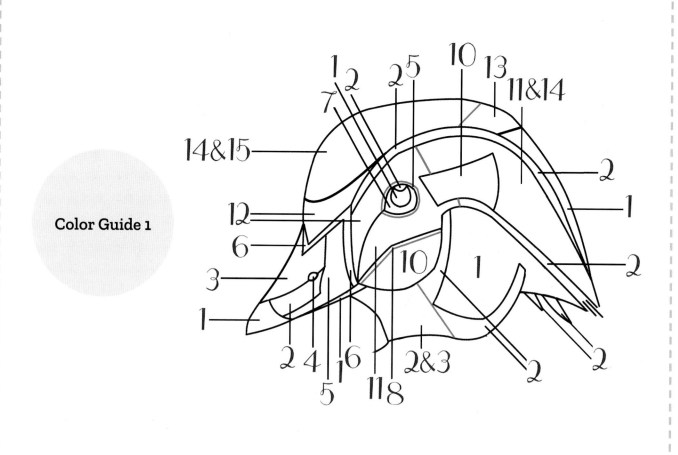

Color Guide 1

STEP 1

Start by stitching the wood duck's beak and eye, referring to Color Guide 1 to help you with colors and details, such as the Coral (351) outline around the eye.

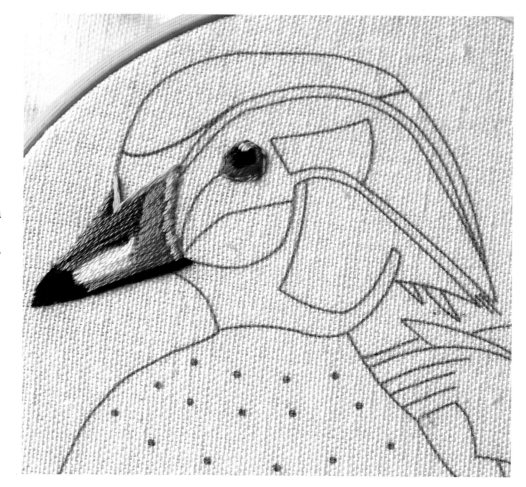

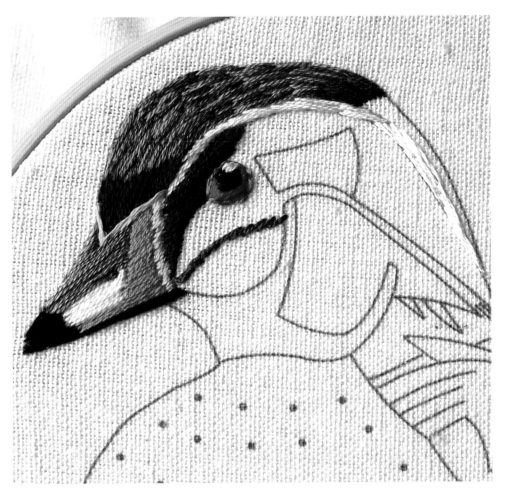

STEP 2

Next, work on the surrounding sections of the wood duck's face. For the Medium Very Dark Cornflower Blue (158) stripe across the wood duck's cheek, stitch with overlapping teeny stitches, following the stitching flow diagram directions.

STEP 3

When stitching the Snow White (B5200) lines across the wood duck's head and face, work around the shape with a few teeny stitches at a time. Several sections on the face require blended stitches; use Color Guide 1 to see where these are placed. Now all the sections of the wood duck's face are stitched!

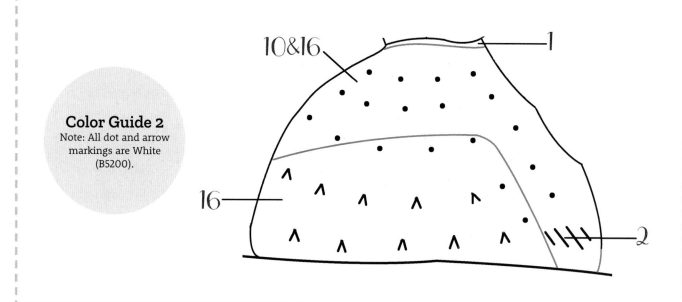

Color Guide 2

Note: All dot and arrow markings are White (B5200).

10&16

1

16

2

STEP 4

Next, stitch the wood duck's chest. Use Color Guide 2 to begin with the thin Black (310) line of stitches just under the wood duck's head before stitching the dotted markings on the upper part of the chest with French knots. (See page 11 for a refresher on French knots.) For the small arrow markings toward the bottom of the chest, create a fan-like series of mini stitches, using the top of the arrow as the point to return to. Also add the Snow White (B5200) dashed markings near the bottom right of the wood duck's belly.

STEP 5

Once you've added the white markings, stitch the sections around them, referring to the gray lines in Color Guide 2 to help you place the colors.

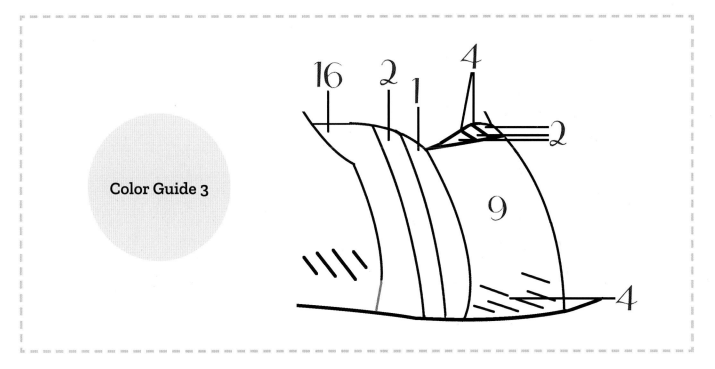

Color Guide 3

STEP 6

It's time to stitch the rest of the chest. Refer to Color Guide 3 as you stitch this part. Start with the three striped sections at the front of the chest and the dashed Very Dark Beaver Gray (645) markings toward the bottom right, under the tail, as well as the small gray striped stitches at the top of the Light Tan (437). Stitch the Light Tan (437) section at the end of the duck's chest and the small Snow White (B5200) stitches between the small Very Dark Beaver Gray (645) stripes.

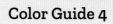

Color Guide 4

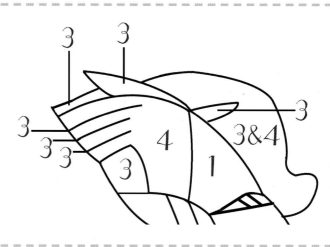

STEP 7

Using Color Guide 4, start with the Light Pewter (169) sections toward the left of the tail. Then add the Very Dark Beaver Gray (645) and Black (310) sections around them.

STEP 8

When you have finished the lower sections of the tail, move on to the final upper tail section. For the final touches, add the Very Dark Gray Green (924) stitches for the water underneath the wood duck. For the water lines, follow the template with a series of short, straight stitches, adjusting the direction as you follow the shape of the lines. You have finished your wood duck—well done for finishing one of the most colorful patterns in this book!

American Flamingo

American flamingos are large wading birds. Their bright pinky-red feathers make them instantly recognizable. The word flamingo has roots in the Spanish word *flamengo*, meaning "flame-colored." Get ready for lots of pinks, reds, and feather details to capture this flamingo's fire-colored hues. I couldn't resist capturing all the tail feathers in this flamingo, so it's a close-up shot.

THREAD COLOR GUIDE				
1 ● DMC 310	4 ● DMC 169	6 ● DMC 355	8 ● DMC 3853	10 ○ DMC 353
2 ○ DMC B5200	5 ● DMC 351	7 ● DMC 720	9 ● DMC 352	11 ● DMC 350
3 ○ DMC 746				

Stitching Flow Directions

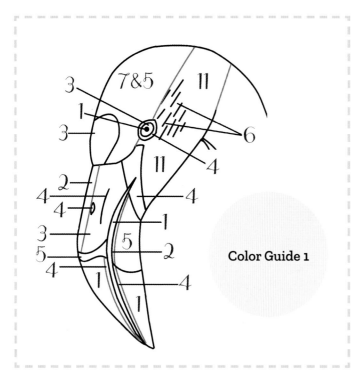

Color Guide 1

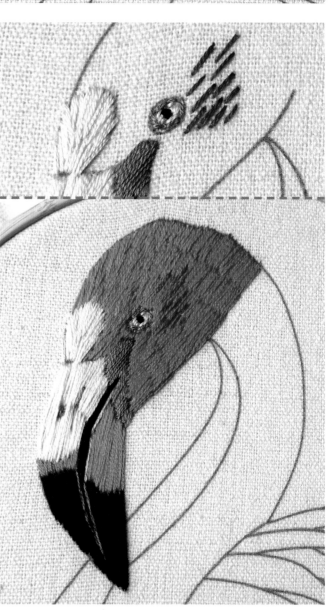

STEP 1

Following Color Guide 1, start by stitching the Light Pewter (169), Snow White (B5200), and Black (310) sections of the flamingo's beak. Note the thinner lines to add along the edges of the Black (310) thin section that curves vertically up the beak. See Color Guide 1 to help you place these thinner lines. You should have one Light Pewter (169) line on the left of the thin, curved black section at the tip of the beak and one on the other side of the central thin, curved section near the end of the beak; these lines are within the Black (310) sections of the beak. The other thin lines are a couple of Snow White (B5200) stitches on the right side of the central black curved section toward the top of the beak. Add the small nostril and curved line in Light Pewter (169).

After these smaller markings, stitch the rest of the beak before stitching the eye and the Off White (746) section at the top of the beak.

STEP 2

Then add the dashed Dark Terra Cotta (355) stitches to the right of the flamingo's eye (see inset) before finishing the sections that make up the flamingo's face.

Color Guide 2

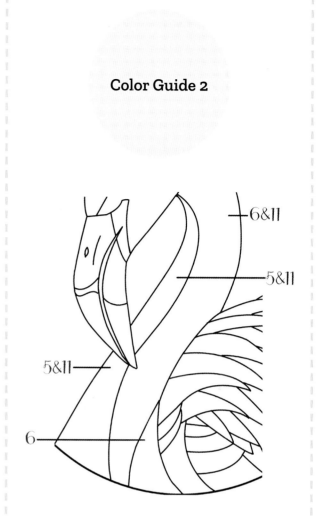

6&11

5&11

5&11

6

STEP 3

Next, work on the flamingo's neck, which is made up of two long curved sections, both with blended stitching. Start with the section on the right of the neck before stitching the section on the left.

STEP 4

Use Color Guide 2 to help you with where to stitch different colors. Then, to finish your flamingo's neck, stitch the smaller Dark Terra Cotta (355) section between the neck and wing.

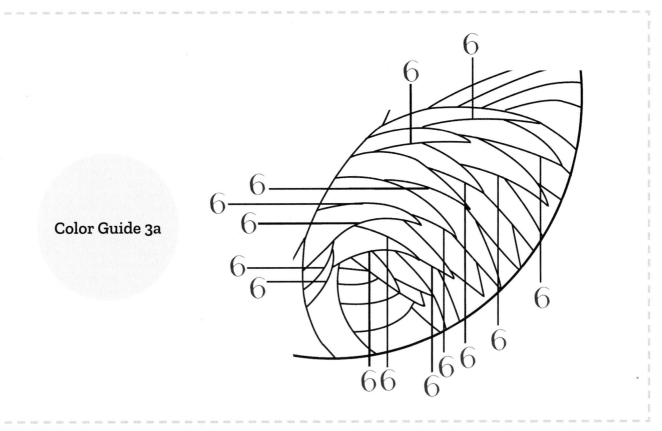

Color Guide 3a

STEP 5

Now you can start the flamingo's curved wing and tail feathers. Start with the several thin lines in Dark Terra Cotta (355). Follow Color Guide 3a to help you place these lines. Follow the lines of the edges of the feathers on your template as you stitch with one to three straight stitches.

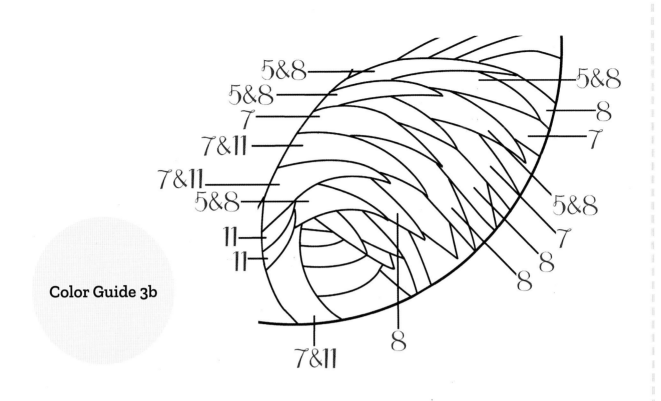

Color Guide 3b

STEP 6

Next, add the upper feather colors to the left of the flamingo's wings, following Color Guide 3b to place the colors. With the feathers you can work on a layer at a time, starting with the upper feathers.

STEP 7

Next, work on the mid-layer of feathers, finishing with the underlayer.

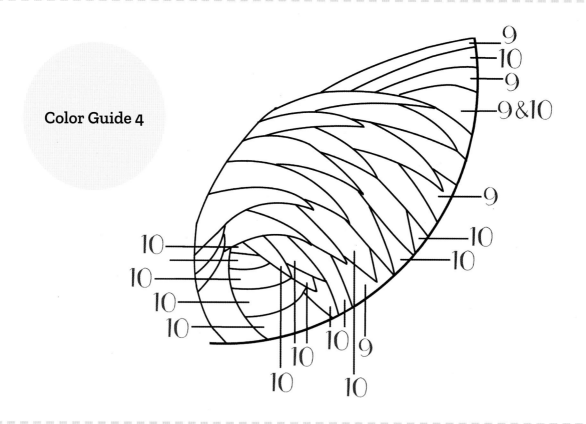

Color Guide 4

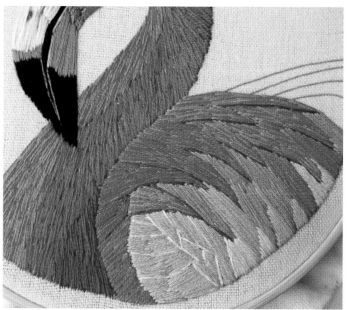

STEP 8

Now it's time to stitch the lighter-colored under feathers that make the second layer of the flamingo's wing and tail feathers, following Color Guide 4. Start with the Snow White (B5200) outlines around the feathers toward the left edge of the wing feathers. (Note: These aren't marked on the Color Guide; follow along with the photo for where to stitch these).

STEP 9

Stitch the main colors of the feathers and the final underlayer of lighter-colored feathers that make up the flamingo's wings. You're on the last bit. Just the tail feathers at the top right of the flamingo left to stitch. For the tail feathers use staged stitching, working around the section shapes.

Kingfisher

This pattern is inspired by the beautiful common kingfisher, also known as the Eurasian kingfisher. Kingfishers are ace fishers, as their name suggests, and are well-designed for super-fast dives into water, often appearing as a flash of turquoise across the water's surface. They are mostly found perching in trees near waterways. This pattern captures an in-flight kingfisher mid-dive, showcasing its bright colors as they flash in the sun.

THREAD COLOR GUIDE				
1 ● DMC 310	4 ● DMC 169	7 ● DMC 437	9 ● DMC 3809	11 ● DMC 3743
2 ○ DMC B5200	5 ● DMC 645	8 ● DMC 3853	10 ● DMC 3846	12 ● DMC 869
3 ● DMC 746	6 ● DMC 720			

Stitching Flow Directions

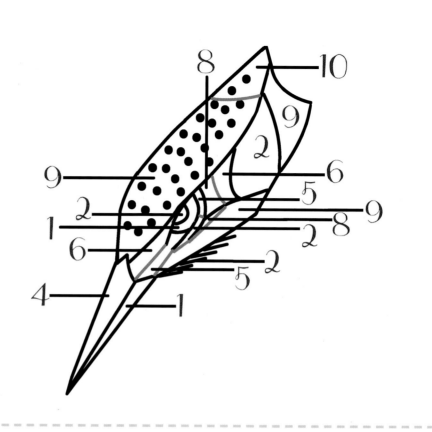

Color Guide 1
Note: All dotted markings are Snow White (B5200).

STEP 1

Start by stitching the kingfisher's beak, eye, and smaller sections around the eye, following Color Guide 1. Use the gray lines in Color Guide 1 to guide you with the extra smaller sections not marked on the pattern template. For the White (B5200) dashed markings on the kingfisher's chin, use a series of small diagonal stitches, following the lines on the pattern template.

STEP 2

Finish the other sections on the lower face before adding French knots over the Snow White (B5200) dotted markings on the kingfisher's head. Then fill in the Ultra Very Dark Turquoise (3808) and Light Bright Turquoise (3846) sections around them.

STEP 3

Next, start stitching the kingfisher's wings, following Color Guide 2 on the following page. Begin with the French knots and turquoise sections on the lower part of the kingfisher's upper wing, using staged stitching (see page 12) for this part of the wings.

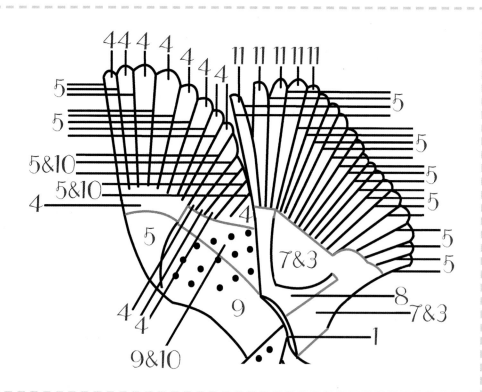

Color Guide 2

Note: All dotted markings are Color No. 2, Snow White (B5200) and all the thin sections between the Color No. 5, Very Dark Beaver Gray (645) stripes on the lower wing are Color No. 11, Very Light Antique Violet (3743).

STEP 4

Add the gray sections on the top-left corner of the upper wing before adding the Very Dark Beaver Gray (645) lines, following the pattern template for the upper part of the wing; for the lower five stripes, add a Light Bright Turquoise (3846) stitch above each one.

STEP 5

Finish the upper wing with the Light Pewter (169) sections between the stripes before moving on to stitching the Very Dark Beaver Gray (645) stripes on the lower wing and the Very Light Antique Violet (3743) sections in between. Finish the orange and gold sections of the kingfisher's lower wing before moving on.

Color Guide 3

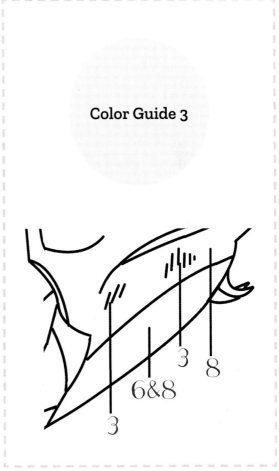

STEP 6

Now that you've finished the kingfisher's wings and face, it's time to stitch the bright-colored chest, following Color Guide 3. Start with the Off White (746) dashed markings on the kingfisher's upper chest section before stitching the Dark Autumn Gold (3853) section around them. Stitch the lower part of the chest with a blend of Dark Orange Spice (720) and Dark Autumn Gold (3853).

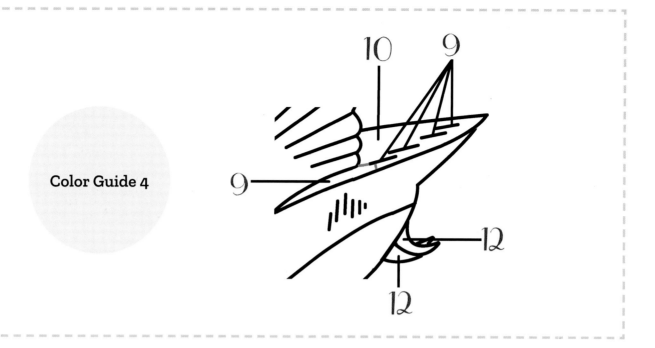

Color Guide 4

STEP 7

There's just the kingfisher's tail and foot to go! Following Color Guide 4, start with the Very Dark Turquoise (3809) markings on the kingfisher's tail before filling in the Light Bright Turquoise (3846) section around it. For the kingfisher's feet (see inset), make a small series of stitches following the template shape in Very Dark Hazelnut Brown (869).

Atlantic Puffin

Atlantic puffins are seabirds found along coastlines of the Atlantic Ocean. Puffins are also expert fishers and are often seen carrying several small fish in their beaks. One of my favorite facts about puffins is that baby puffins are called pufflings.

This pattern is for a puffin portrait, so you can see the lovely beak, as well as the recognizable bright, orangey-red feet.

THREAD COLOR GUIDE				
1 ● DMC 310	3 ○ DMC 746	5 ● DMC 169	7 ● DMC 840	9 ● DMC 743
2 ○ DMC B5200	4 ● DMC 645	6 ● DMC 720	8 ● DMC 666	

Stitching Flow Directions

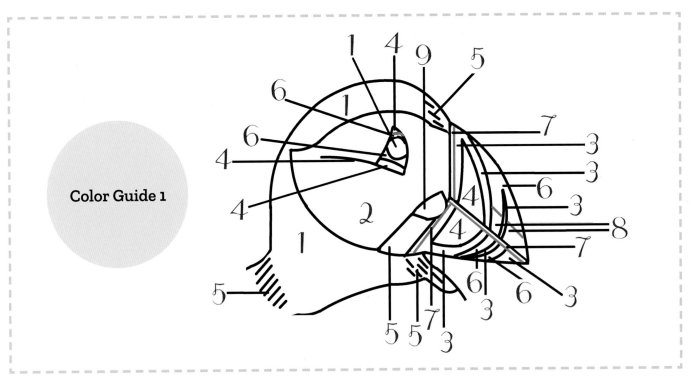

Color Guide 1

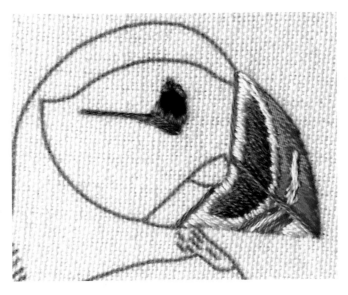

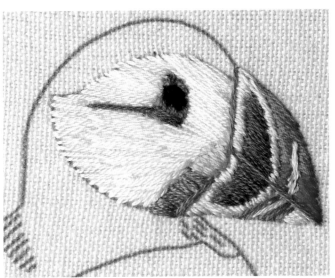

STEP 1

Start by stitching the puffin's eye and beak, following Color Guide 1. For some of the thin sections, such as Off White (746) and Dark Orange Spice (720) on the underside of the beak, you can go against the stitching flow directions diagram where easier. For the part of the eye above the Black (310) center of the puffin's eye, add just two tiny stitches as a thin upper outline for the pupil, using Dark Orange Spice (720). Add the thin lines of Medium Beige Brown (840) in the puffin's beak last. One of these lines follows the center line of the beak, and the other two follow the lines at the left edge of the beak. Use the gray lines shown here and in Color Guide 1 to help you place these.

STEP 2

When the beak and eye are done, stitch the smaller sections of Medium Yellow (743) and Light Pewter (169) next to the puffin's beak, before adding the larger Snow White (B5200) section around the eye. Add the Light Pewter (169) dashed markings at the top of the puffin's head, under the chin, and on top of the shoulder.

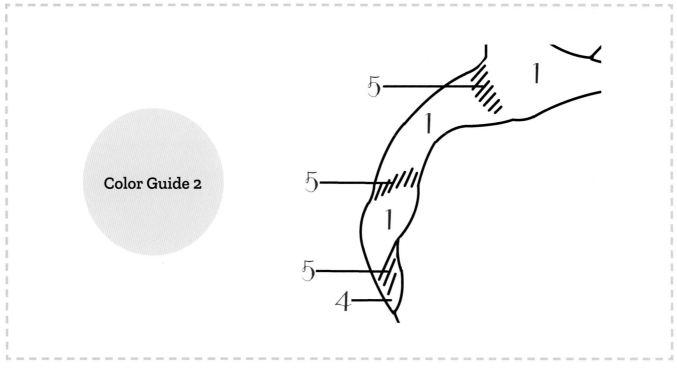

Color Guide 2

STEP 3
Stitch the larger Black (310) section that makes up the outer side of the puffin's head and the start of the neck. For the larger Snow White (B5200) and Black (310) sections use staged stitching.

STEP 4
Using Color Guide 2, move on to stitching the other two areas of Light Pewter (169) dashed markings toward the bottom of the puffin's wings.

STEP 5

Stitch the rest of the Black (310) section that makes up the puffin's back with staged stitching and the small Very Dark Beaver Gray (645) section toward the bottom of the Black (310) section. At this point, you can also add the Light Pewter (169) dashed markings toward the bottom of the puffin's chest.

Color Guide 3

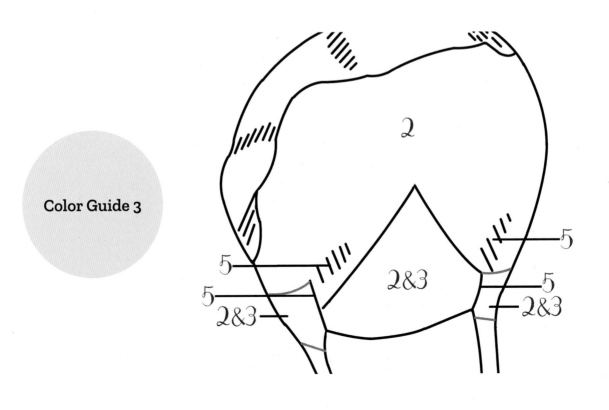

STEP 6

Next you can start stitching the puffin's chest. Begin with the two Light Pewter lines that define the inner edges of the puffin's legs. Then stitch the lower triangular section of the puffin's chest using blended and staged stitching. Use Color Guide 3 to guide you with colors placement as you stitch.

STEP 7

Then move on to stitching the upper part of the puffin's chest. For this larger Snow White (B5200) section you can scatter your stitches (see page 13), creating one layer of feather-like flow before adding another layer. Finally, add the stitches that make up the fluffy Snow White (B5200) and Off White (746) tops of the puffin's legs.

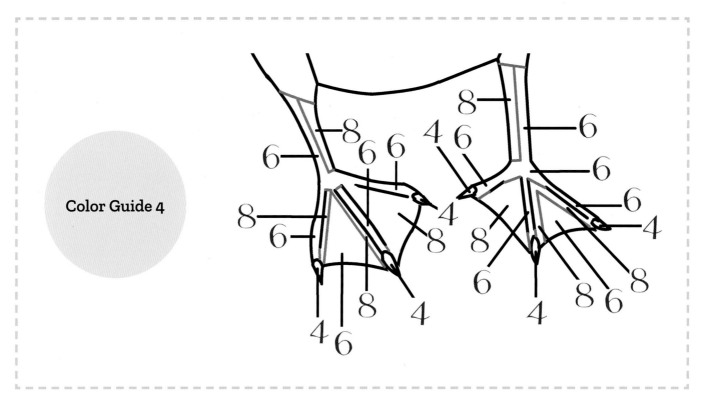

Color Guide 4

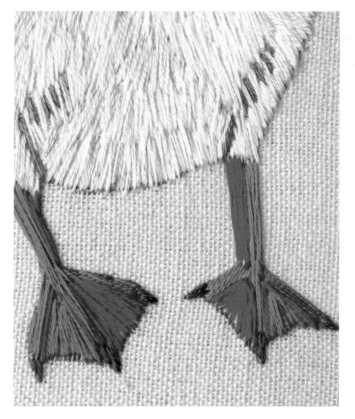

STEP 9

You're nearly done stitching your puffin! Using Color Guide 4, stitch the puffin's feet, starting with the Bright Red (666) sections.

STEP 10

Then stitch the Dark Orange Spice (720) sections in between the Bright Red sections. As you stitch the feet, imagine a puffin's bone structure as a guide for where to place the lighter and darker colors. The claws are made up of just one to three teeny Very Dark Beaver Gray (645) stitches each.

Swallow

There are many species of swallows around the world. Their pointed tails and swift flight make them recognizable in the air. Historically, sailors often got swallow tattoos as marks of experience and their ability to survive difficult journeys and return home. This pattern captures a swallow in flight, when its strong wings and pointed tail can be seen at their best.

THREAD COLOR GUIDE						
1 DMC 310	3 DMC 746	5 DMC 351	6 DMC 645	7 DMC 355		
2 DMC 924	4 DMC 169					

Stitching Flow Directions

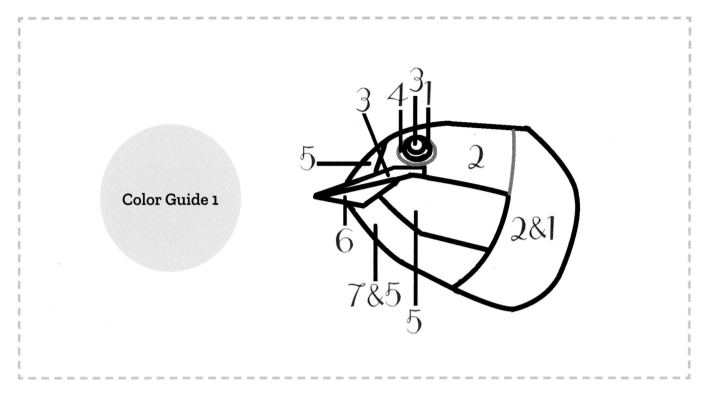

Color Guide 1

STEP 1

Start by stitching the Coral (351) and Dark Terra Cotta (355) sections of the swallow's chin and the teeny sections of Very Dark Beaver Gray (645) and Off White (746) that make up the beak. The Off White (746) line on the beak is just a single stitch. Then work on the swallow's tiny eye, referring to Color Guide 1 for color placement. Start with the Off White (746) circle in the middle; if one teeny stitch doesn't stand out enough, add another teeny stitch on top of the first. Then stitch the surrounding Black (310) section. Next, add a small Light Pewter (169) outline just under the bottom off the eye and around each side with three teeny stitches, following the circular shape. When the eye is done, stitch the Very Dark Grey Green (924) and Black (310) sections of the swallow's face.

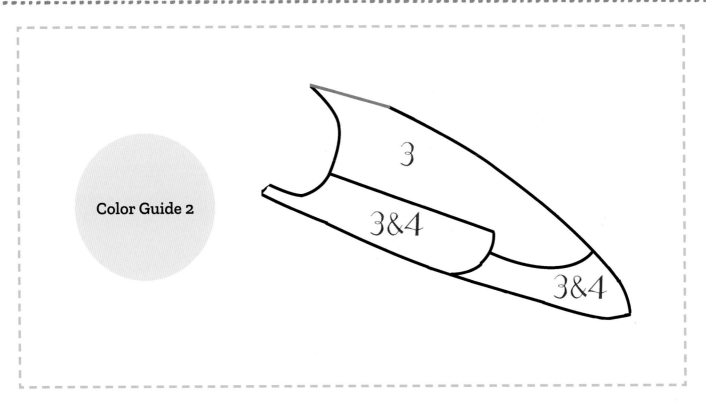

Color Guide 2

3

3&4

3&4

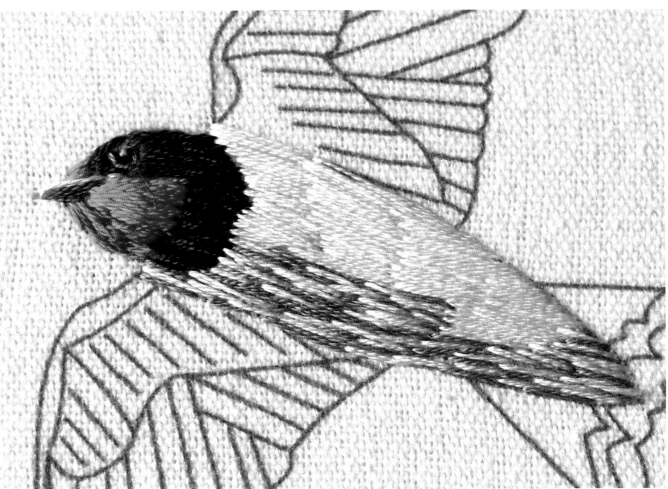

STEP 2

Now for an easy bit! Move on to the swallow's chest, which is made up of three sections. Refer to Color Guide 2 for the color placement.

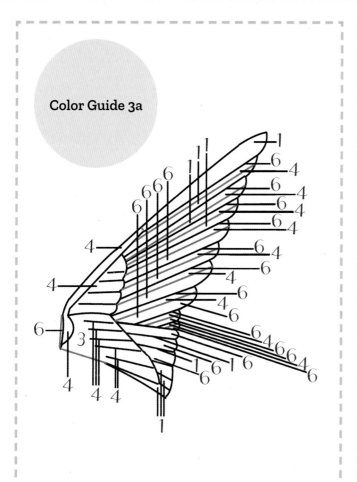

Color Guide 3a

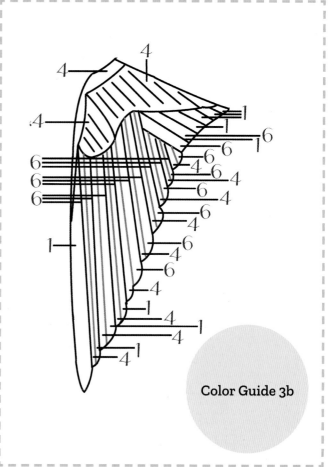

Color Guide 3b

STEP 3

Next, stitch the wings, referring to Color Guide 3a, which shows the upper wing, and Color Guide 3b, which shows the lower wing. It may look like a lot of numbers and lines at first, but just take each section one at a time and you'll be done before you know it! Start with the Very Dark Beaver Gray (645) lines on the swallow's outer wings and shoulders, following the pattern template.

STEP 4

Then add two Light Pewter (169) stitches on the inside of each Very Dark Beaver Gray (645) wing line stitch and the Light Pewter (169) sections on the swallow's shoulders.

STEP 5

Next, add the Light Pewter (169) sections and stripes on the inner wings.

STEP 6

Add the Black (310) stripes and sections. For the final touches on the wings, stitch the Very Dark Beaver Gray (645), Light Pewter (169), and Off White (746) parts of the wings (see inset).

Color Guide 4

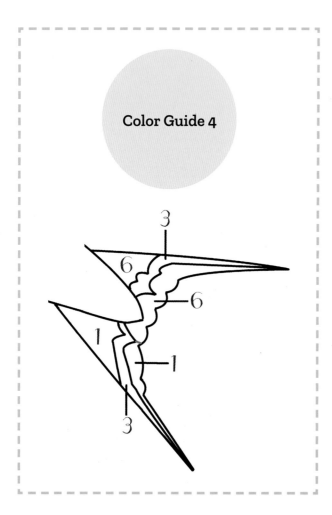

STEP 7

You're nearly there! Just the swallow's tail left. Start by stitching the Black (310) and Very Dark Beaver Gray (645) sections on either side of the tail. Then stitch the final Off White (746) sections to finish your swallow pattern!

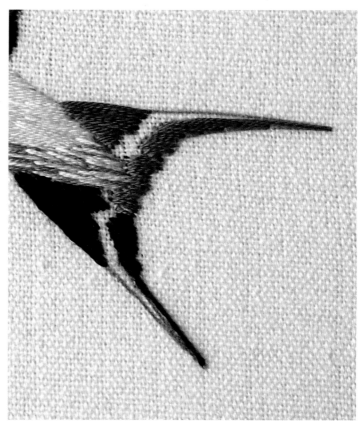

Barn Owl

Barn owls are beautiful nocturnal creatures often found in barns, as their name suggests. They can appear ghostly as they fly and hunt at night, with their creamy white bellies shining in the dark. This portrait pattern captures lots of feather markings, so it may feel fiddly in parts, but it's so worth it to show off this barn owl's lovely, dotted details.

THREAD COLOR GUIDE					
1 ● DMC 310	4 ● DMC 645	6 ● DMC 3853	8 ● DMC 437	10 ● DMC 300	
2 ○ DMC B5200	5 ● DMC 169	7 ● DMC 783	9 ● DMC 840	11 ● DMC 778	
3 ○ DMC 746					

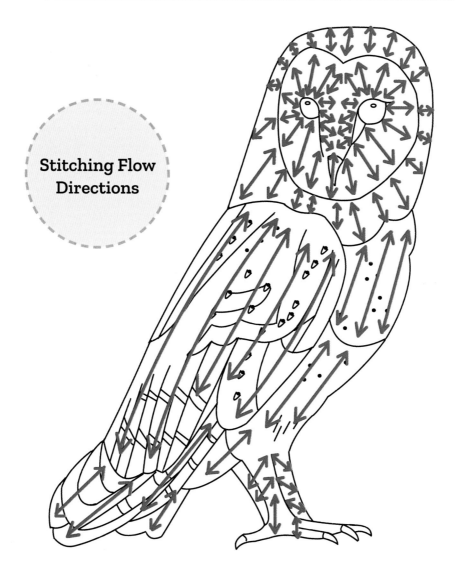

Stitching Flow Directions

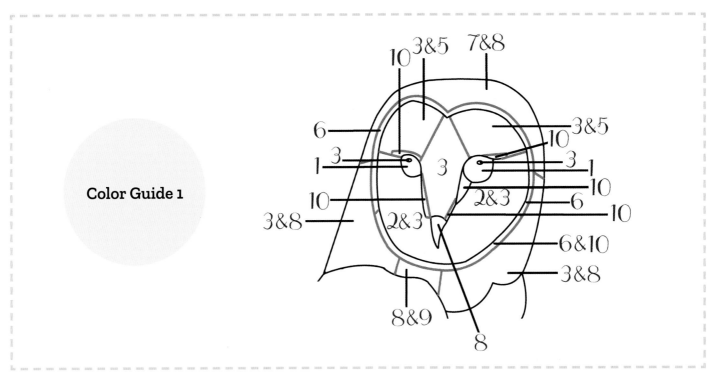

Color Guide 1

STEP 1

Referring to Color Guide 1, start by stitching the owl's eyes and the Very Dark Mahogany (300) lines around the eyes.

STEP 2

Next, stitch the rest of the sections that make up the owl's face and beak. For the central section that makes up the nose and forehead, imagine a line down the middle of the section and work outward in a fanlike series of stitches, using the Stitching Flow Diagram on page 83 to guide you. Finish the face by adding the Dark Autumn Gold (3853) and Very Dark Mahogany (300) in thin, circular sections around the owl's face with rows of teeny stitches, referring to the gray lines in Color Guide 1 for the placement of these colors and the other smaller sections that aren't marked on your pattern template.

STEP 3

When you've finished the face, stitch the head sections around it.

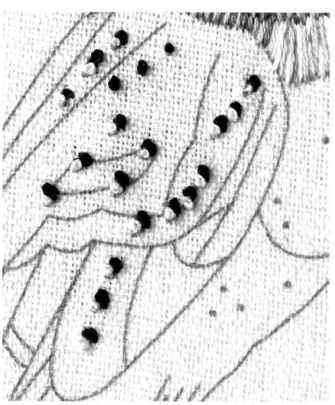

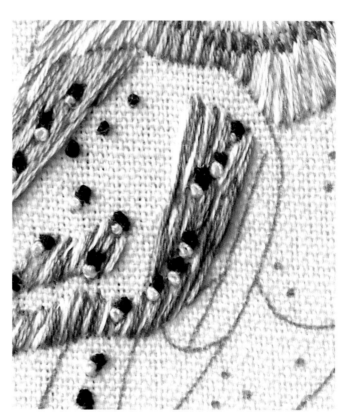

STEP 4

Next, work on the dotted markings across the owl's wings. For the larger black-and-white markings, stitch one Black (310) French knot at the top of each marking (where the black dot is) and a Snow White (B5200) French knot underneath. Use single French knots for the three Black (310) single dotted markings to the top left of the owl's wing.

Now stitch the sections to the left of the upper wing and the Very Dark Beaver Gray (645) and Off White (746) parts of the wing, referring to Color Guide 2 on the following page to place the colors. Then stitch the Dark Autumn Gold (3853) and Medium Topaz (783) sections around them.

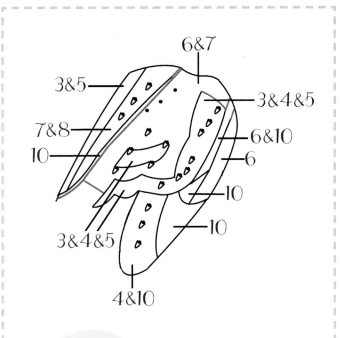

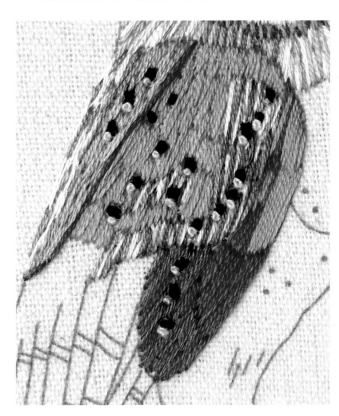

Color Guide 2

Note: All the small two-part oval wing markings are Black (310) on top and Snow White (B5200) on the bottom. All single-dotted markings are Black (310).

STEP 5

Finish the owl's upper wings with the Very Dark Mahogany (300) and Very Dark Beaver Gray (645) sections toward the bottom of the upper wings.

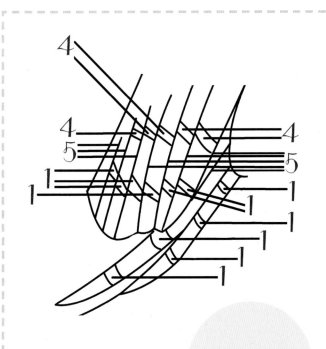

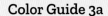

Color Guide 3a

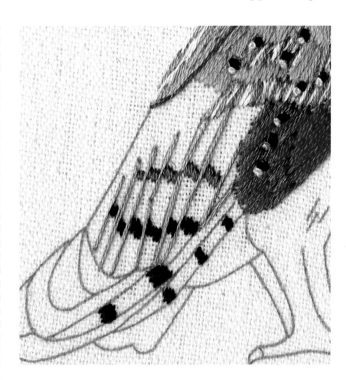

STEP 6

Using Color Guide 3a, start stitching the owl's lower wings with the Light Pewter (169) wing lines and the small Black (310) and Very Dark Beaver Gray (645) stripes across the feathers.

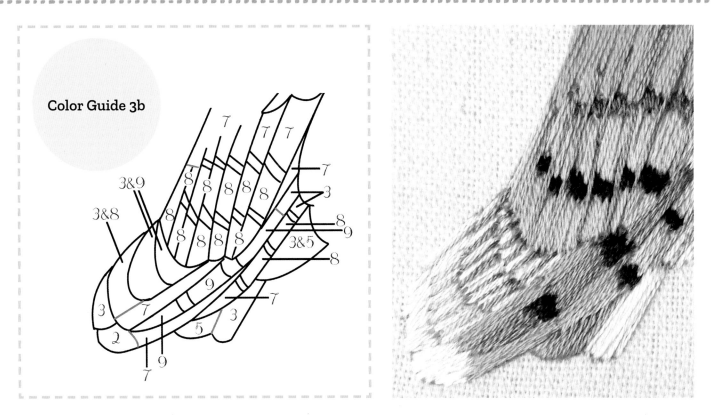

STEP 7

Refer to Color Guide 3b as you work your way down the owl's lower wing feathers, stitching the Medium Topaz (783) and Light Tan (437) sections and finishing with the blended sections and Off White (746) sections toward the end of your owl's tail. Then stitch the inner feather section to the right of the owl's lower wings.

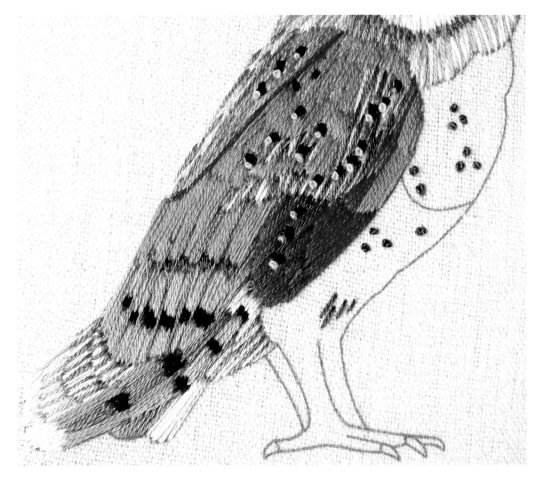

STEP 8

You're nearly there—just the last few sections of the owl's chest and legs to go. Start by stitching the dashed lines above the legs and the dotted markings on the chest with French knots in Very Dark Beaver Gray (645).

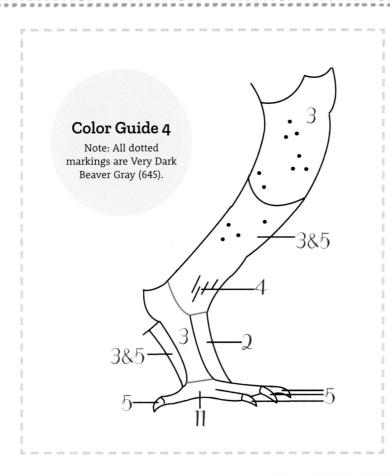

Color Guide 4

Note: All dotted markings are Very Dark Beaver Gray (645).

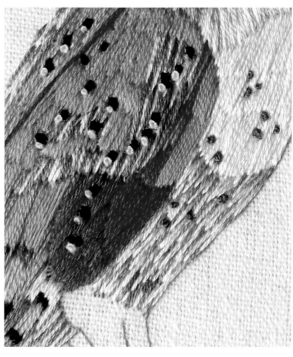

STEP 9

Next, stitch the sections around the dotted and dashed markings with staged stitching, referring to Color Guide 4.

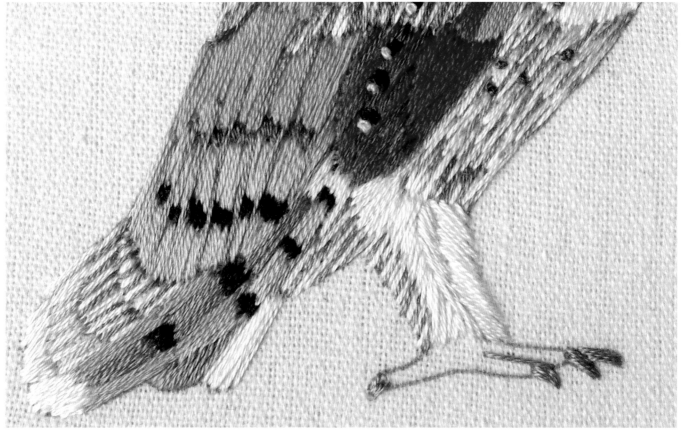

STEP 10

Finish with the fluffy legs and feet. Stitch each claw with just a couple teeny, snug stitches. Stitch the claws before stitching the Very Light Antique Mauve (778) sections that make up the feet.

Hoopoe

Hoopoes are distinctive birds found across Asia, Europe, and Africa. They are recognizable by the distinctive orange-and-black dotted crest across the top of their heads. This pattern zooms in on the hoopoe, capturing its crown in all its finery.

THREAD COLOR GUIDE				
1 DMC 310	4 DMC 169	7 DMC 720	10 DMC 437	13 DMC 164
2 DMC B5200	5 DMC 402	8 DMC 3853	11 DMC 3778	14 DMC 869
3 DMC 746	6 DMC 743	9 DMC 783	12 DMC 840	15 DMC 645

Stitching Flow Directions

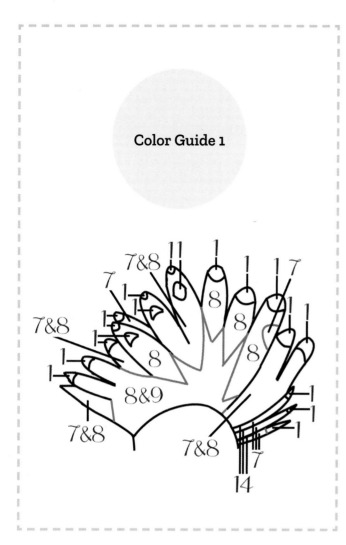

Color Guide 1

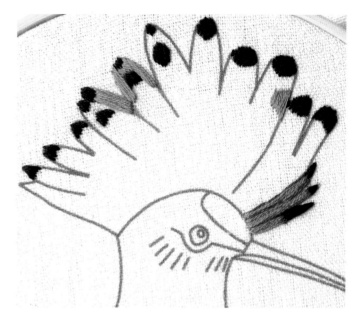

STEP 1

Using Color Guide 1, start by stitching the head feathers of the hoopoe's crest. It may help to add the gray lines marked on Color Guide 1 onto your fabric template with a pencil or pen to guide you as you stitch. Don't worry about getting these lines exact. It's easiest to begin with the Very Dark Hazelnut Brown (869) lines that outline the thin sections to the right of the crest, and the smaller sections of Black (310) and Dark Orange Spice (720).

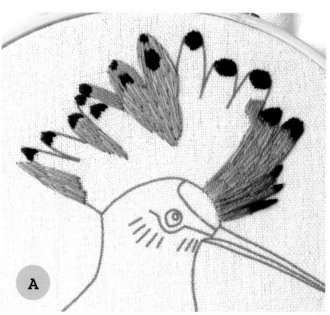

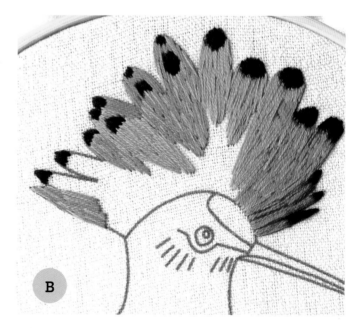

STEP 2

Refer to the gray lines in Color Guide 1 to add the blended sections of Dark Orange Spice (720) and Dark Autumn Gold (3853) (see image A) before working on the Dark Autumn Gold (3853) feathers (see image B). Finish the crest by stitching the blended Dark Autumn Gold (3853) and Medium Topaz (783) section between the feathers.

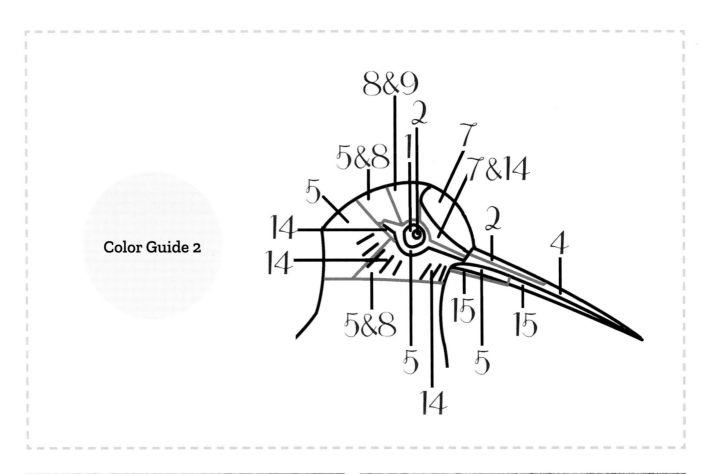

Color Guide 2

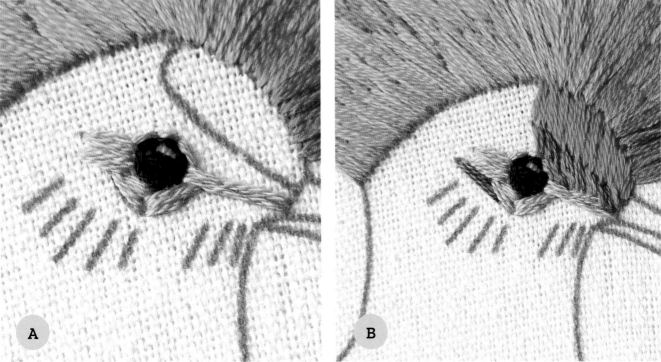

A

B

STEP 3

Move on to the face and beak, referring to Color Guide 2. Start with the small Very Light Mahogany (402) section around the hoopoe's eye (A). Once you've stitched the first section around the eye, stitch the Very Dark Hazelnut Brown (869) and Dark Orange Spice (720) sections above the eye, as well as the Very Dark Hazelnut Brown (869) line under the first section around the eye (B).

STEP 4

Next, stitch the Very Dark Hazelnut Brown (869) dashed lines under the eye before adding the sections around them.

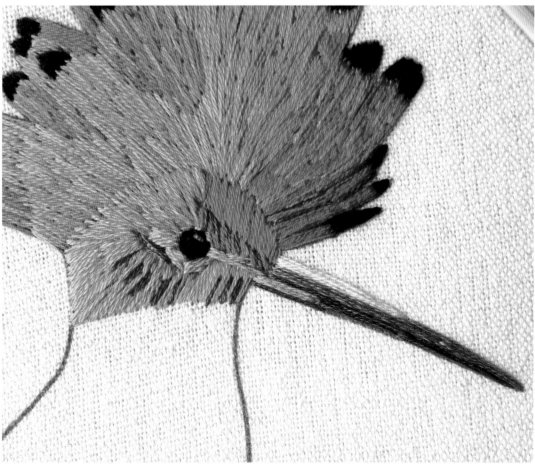

STEP 5

Refer to Color Guide 2 for color placement as you stitch the hoopoe's beak. Each beak section needs only one to three long thin stitches.

Color Guide 3

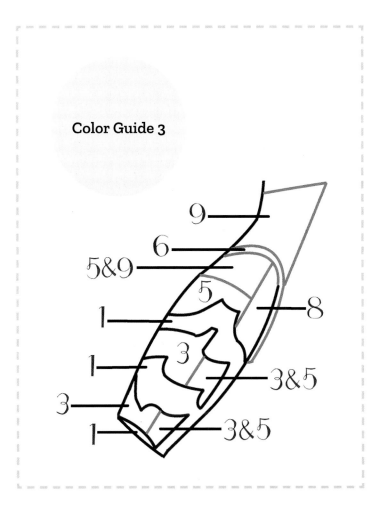

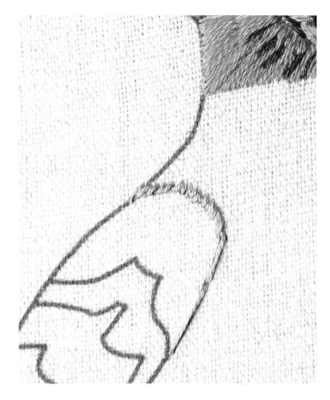

STEP 6

Using Color Guide 3, start by stitching the thin Medium Yellow (743) line around the top of the hoopoe's wing.

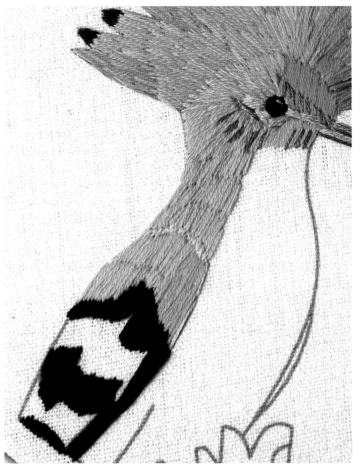

STEP 7

Next, add the Medium Topaz (783) section at the back of the hoopoe's neck before working on the sections of the hoopoe's wing, starting at the top of the wing. Stitch the Black (310) sections before filling in the Off White (746) and Very Light Mahogany (402) sections that make up the rest of the wing.

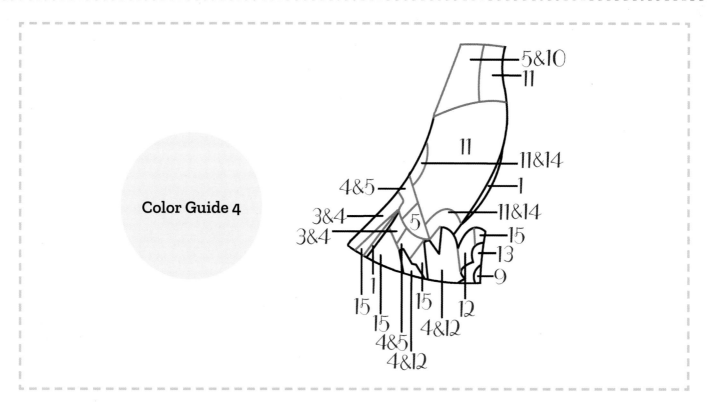

Color Guide 4

STEP 8

It's time to start stitching the hoopoe's chest. Begin at the top before working on the Light Terra Cotta (3778) and Very Dark Hazelnut Brown (869) sections in the middle, using staged stitching (see page 12).

STEP 9

Use the gray lines in Color Guide 4 to help place the colors at the bottom of the hoopoe's chest. Finish with the branch the hoopoe sits on, working from left to right, finishing with the Light Forest Green (164) and Medium Topaz (783) sections. Stitch these last two sections with French knots, stitching them closely together to cover the whole of each section.

Raven

Found on several continents, ravens have rich symbolism in folklore in many cultures. Ravens are curious birds and are often seen strutting from place to place, making their presence known through their distinctive, hearty croak-like call. Unlike their corvid cousins, crows, ravens tend to travel in pairs rather than large groups. With this pattern, you'll get up close to this inquisitive bird.

THREAD COLOR GUIDE				
1 ● DMC 310	**2** ○ DMC B5200	**3** ● DMC 869	**4** ● DMC 169	**5** ● DMC 645

Stitching Flow
Directions

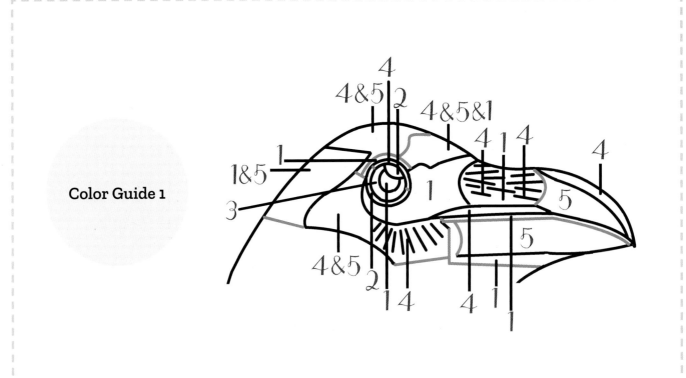

Color Guide 1

STEP 1

Start by stitching the raven's eye, using Color Guide 1 to place the colors (see inset). Stitch the Black (310) section on the raven's face, and then add the

Light Pewter (169) dashed lines under the eye and across the top of the beak before working on the other sections of the beak. At this point, you can also add the Black (310) outline around the edge of your raven's eye.

STEP 2

Stitch the rest of the sections of the raven's face.

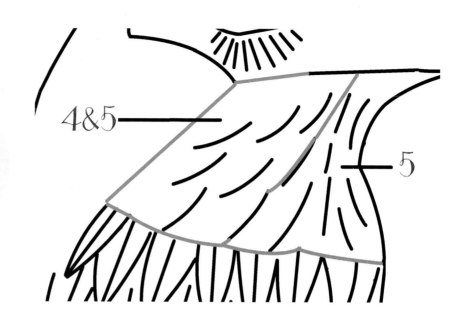

Color Guide 2

4 & 5

5

Note: All curved line markings are Very Dark Beaver Gray (645).

STEP 3

Stitch the raven's neck, referring to Color Guide 2, starting with the Very Dark Beaver Gray (645) dashed markings on your template. These can be made with one or two straight stitches covering the lines.

STEP 4

Use the markings you stitched in the previous step as a guide for the stitching flow as you stitch the sections around them, starting with the right side and finishing with the left.

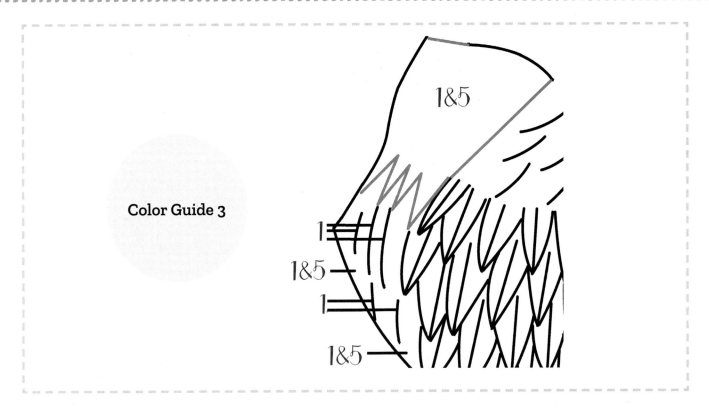

Color Guide 3

1&5

1

1&5

1

1&5

STEP 5

Next, work on the raven's upper and lower back, referring to Color Guide 3. Start with the upper back, using staged stitching to stitch the Black (310) stitches before blending in the Very Dark Beaver Gray (645) stitches. Create some points toward the bottom of this section to resemble feather ends, using an arrow-like formation of stitches. Note the gray lines in Color Guide 3 to help with this part.

STEP 6

Then fill in the rest of this section with the blended Black (310) and Very Dark Beaver Gray (645) stitches.

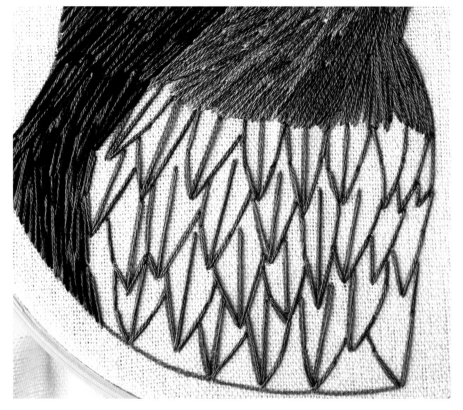

STEP 7

You're almost done! Time to stitch the raven's dapper chest feathers. Start by stitching the Very Dark Beaver Gray (645) outlines and central lines for each chest feather on your pattern template. For curved lines, use two to three short straight stitches, following the shape of the template; for straight lines, use a single straight stitch. Then stitch the Very Dark Beaver Gray (645) sections of the feathers on the left, referring to Color Guide 4.

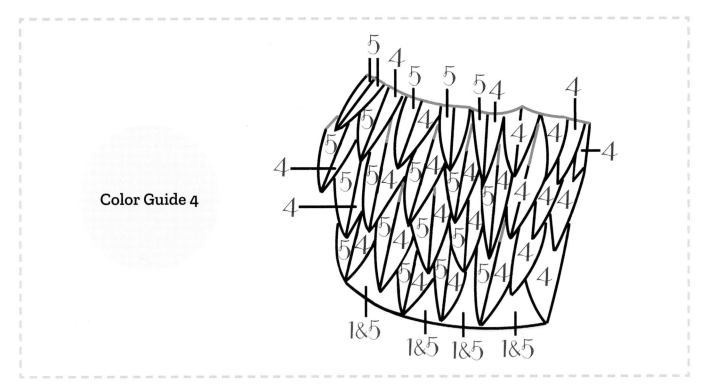

Color Guide 4

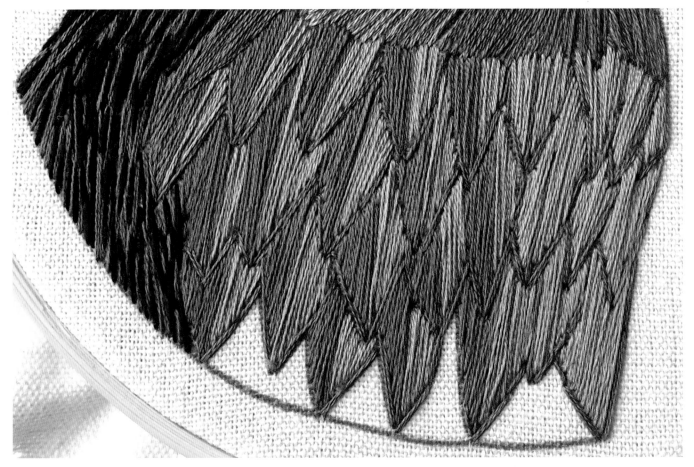

STEP 8

Now stitch the Light Pewter (169) stitches that make up the rest of the raven's chest feathers before finishing with the blended Black (310) and Very Dark Beaver Gray (645) sections toward the bottom of the chest.

Eastern Bluebird

Recognizable by their vivid blue feathers and orange-colored bellies, eastern bluebirds are found across eastern North America. These blue beauties can be spotted in grasslands and around trees in open country. This pattern captures an eastern bluebird in flight, with lots of wing details.

	THREAD COLOR GUIDE			
1 DMC 310	4 DMC 645	7 DMC 996	9 DMC 3776	11 DMC 157
2 DMC B5200	5 DMC 169	8 DMC 3846	10 DMC 926	12 DMC 3778
3 DMC 746	6 DMC 783			

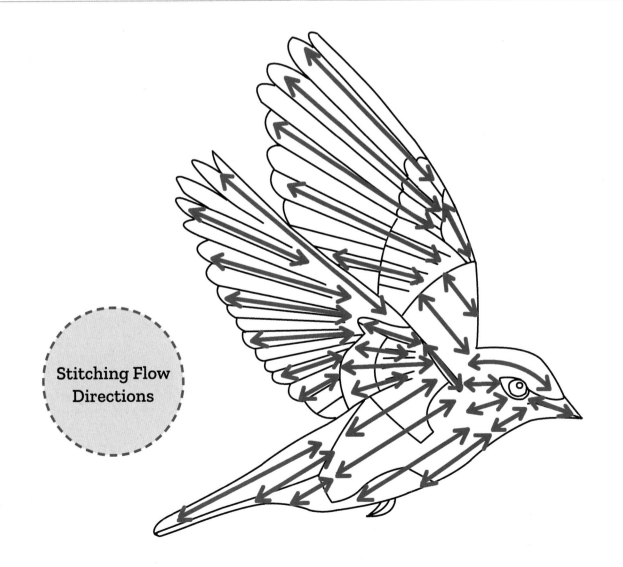

Stitching Flow Directions

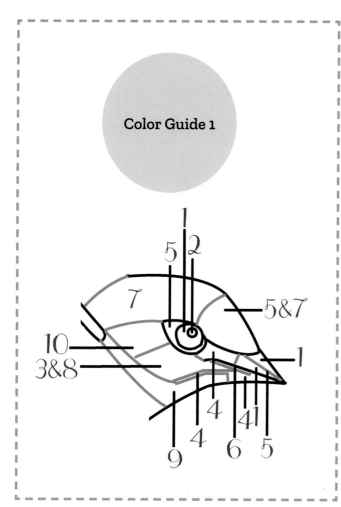

Color Guide 1

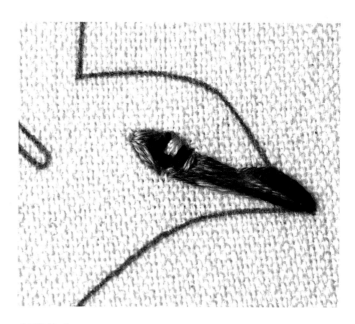

STEP 1

Start by stitching the bluebird's eye and beak, referring to Color Guide 1. Begin with the smallest sections around the eye, using the gray lines in Color Guide 1 to guide you. There are a couple of single-stitch details, shown in Color Guide 1, around the beak: one in Medium Topaz (783) that follows the mouth and one Light Pewter (169) line halfway across the top of the beak.

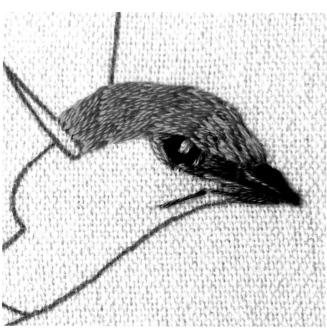

STEP 2

Next, stitch the Medium Electric Blue (996) and Light Pewter (169) sections on the top of the bluebird's head, followed by the Very Dark Beaver Gray (645) line under the beak and the bottom of the beak.

STEP 3

Finish the bluebird's face with the Light Mahogany (3776) section of the chin.

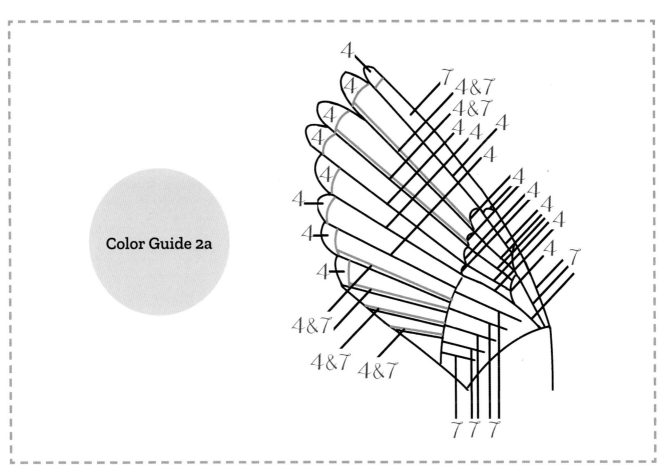

Color Guide 2a

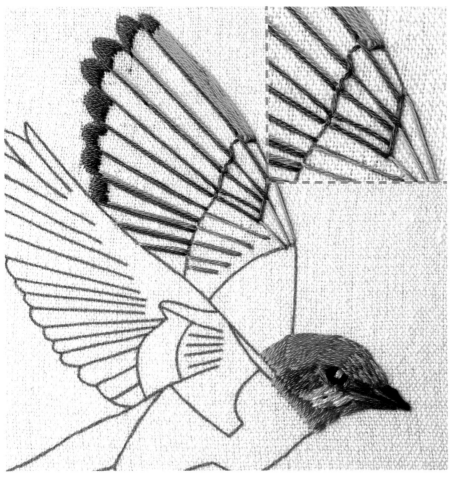

STEP 4

Moving to the bluebird's upper wing, start with the wing lines across the edges of the feathers and feather details at the tips in Very Dark Beaver Gray (645), as well as the wing lines in Medium Electric Blue (996), using Color Guide 2 for color placement. Note the tiny Very Dark Beaver Gray (645) stitches in the pattern template at the top and bottom of the smaller wing feathers in the upper-middle right of the upper wing (see inset). Use one or two teeny stitches for each of these feathers.

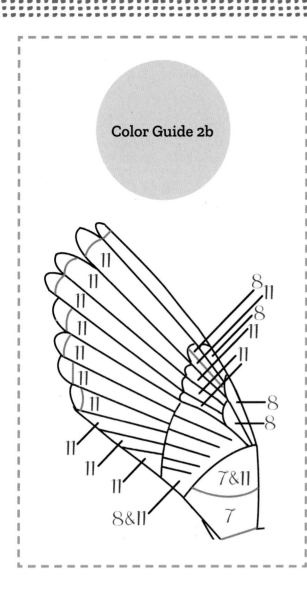

Color Guide 2b

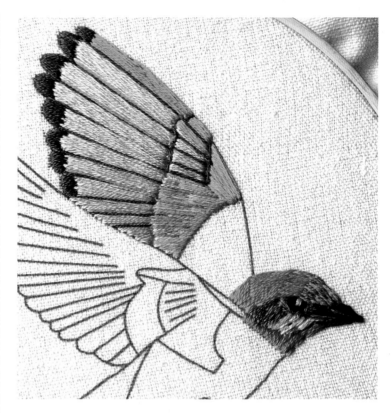

STEP 5

Next, stitch the Very Light Cornflower Blue (157) stitches in between the lines on the upper wing, and the Light Bright Turquoise (3846) sections that make up the rest of the upper wing feathers. Finish the upper wing with the Medium Electric Blue (996) and Very Light Cornflower Blue (157) section at the bottom of the upper wing.

STEP 6

You're ready to stitch the bluebird's lower wing. Referring to Color Guide 3 on the following page, start by stitching the Very Dark Beaver Gray (645) and Light Pewter (169) wing lines and wing details on the tips and upper edge of the wing feathers.

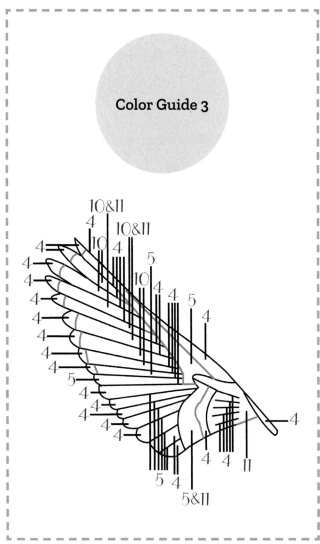

Color Guide 3

STEP 7

Then add the stitches in between the wing lines and the wing details in the lower part of the wing.

STEP 8

Finish the bluebird's lower wing with the Very Light Cornflower Blue (157) section toward the bottom of the wing.

STEP 9

You are nearly there! Just the chest and tail to stitch. Start by stitching the sections around the bluebird's chest.

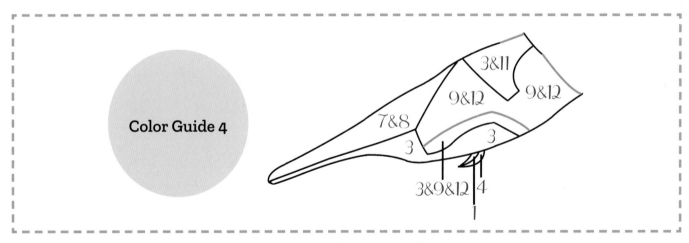

Color Guide 4

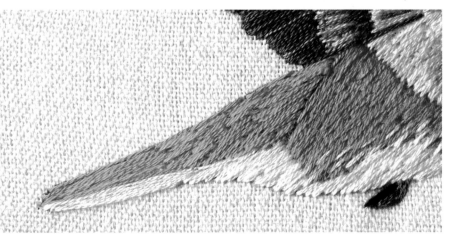

STEP 10

Finish your bluebird with the tail, referring to Color Guide 4 to guide you.

PATTERN
Templates

Calliope Hummingbird

Cardinal

Keen-Billed Toucan

Wood Duck

American Flamingo

Kingfisher

Atlantic Puffin

Swallow

Barn Owl

Hoopoe

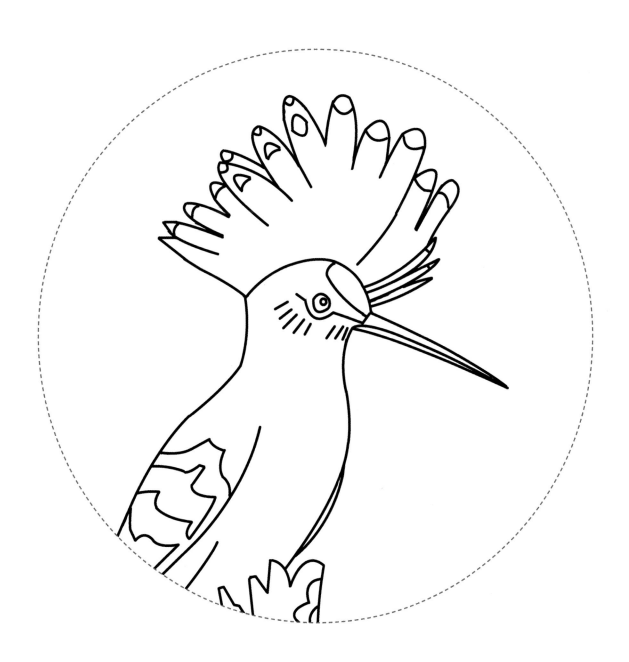

Raven

Eastern Bluebird

About the Author

Beth Hoyes is a British-American embroidery artist and art therapist. Based in Denver, Colorado, Beth has a love of art and nature, as well as great appreciation for wildlife and conservation. With a background as an art therapist, Beth believes in the power and soothingness of making and loves to share this with others. Rabbit Hat Designs is the small embroidery business that Beth runs from her home in Denver, and it focuses on celebrating creativity and nature through her own artwork and accessible embroidery patterns and kits. When not stitching, Beth is spending time with her husband and son in Denver and the Rocky Mountains.

ALSO AVAILABLE FROM WALTER FOSTER PUBLISHING

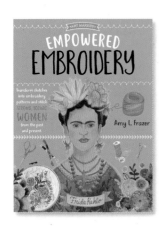

**Art Makers:
Empowered Embroidery**
978-1-63322-884-9

**Art Makers:
Papier Mache**
978-1-63322-892-4

**Cross Stitch Celebrations:
Bundle of Joy!**
978-0-7603-7538-9

Visit www.WalterFoster.com